iPhone 14

USER GUIDE

An Easy & Exhaustive Step-By-Step Manual For Non-Tech-Savvy.

Discover The Best Tips & Tricks To Get The Max Out Of

Your New IPhone 14 In No Time!

VIN MAYER

TABLE OF CONTENTS

HISTORY OF IPHONE

A pple's iPhone is one of the most successful and iconic products in the history of technology. It revolutionized how people interacted with their phones and ushered in a new era of mobile computing. The first iPhone was released in 2007, and Apple has continued to innovate with each new model.

The iPhone has undergone several iterations over the years, from its introduction of revolutionary features like multi-touch, a virtual keyboard, and an App Store to more recent advancements such as facial recognition technology and augmented reality. Each new version of the phone has been met with excitement and criticism from consumers worldwide.

This TOPIC will explore the history of Apple's iPhone, from its humble beginnings to its current standing as one of the most popular devices on the market today. We will examine how it changed our lives, impacting other industries, and what we can expect from future iPhones.

June 2007

In June 2007, the first generation of the iPhone came out in the United States. Steve Jobs had earlier announced the first iPhone in January 2007. It was a combination of the iPod, a new kind of cell phone, and a new way to communicate over the Internet. It had a 3.5-inch screen, a multi-touch touchscreen, a microphone, controls for the headset, and other features that smartphone users today take for granted.

The Wall Street Journal said at the time that the phone was "on balance, a beautiful and innovative handheld computer." TIME Magazine went even further and called the first iPhone the "Invention of the Year."

July 2008

The first phone to beat the iPhone" was released in July 2008. The iPhone 3G came out a year after the first iPhone. It replaced the first iPhone. The second-generation device had some new hardware features, like 3G data and GPS, but the App Store may have been the most important. Users can browse and download third-party apps from Apple's supply stage which, as of early 2018, had more than 2 million apps.

The WSJ said that the iPhone 3G fixed the two main problems with the original iPhone: the price and the fact that it couldn't connect to the fastest cell phone networks. The article said, "The iPhone has a chance to become a real computing platform with a lot of uses."

June 2010

In June 2010, "This changes everything again." The first iPhone that could be used on Verizon's wireless network had a lot of groundbreaking features, all of which would be found in the iPhone 4. Some of these were a high-resolution Retina Display, Apple's first front-facing camera, the ability to do more than one thing at once, and Face Time.

Engadget said that the iPhone 4 is the best smartphone on the market and that "when it comes to the whole package, including fit and finish in both software and hardware, performance, app selection, and all the little things that make a device like this what it is, we think it's the cream of the crop."

The iPhone 4 would have the longest life of any iPhone, and a white version came out less than a year after it came out.

October 2011

October 2011: Saying goodbye to Steve Jobs and hello to Siri. The co-founder of Apple died the day after the iPhone 4s was announced, but he was in charge of making the iPhone 5, which went on sale in September 2012. The "s" stood for Siri, a smart personal assistant that was only available on the 4s at the time.

In addition to Siri, the best-selling iPhone ever had an operating system called iOS 5 that added iMessage, iCloud, and the Notification Center, among other features. Apple's first 8-megapixel camera that could record 1080p video was also on the iPhone 4s.

The Verge called the 4s "a beautiful device with impressive specs," called Siri "one of the most unique apps Apple has ever made," and said, "If this were a car, it would be a Mercedes."

September 2014

Plus-size models came out in September 2014. The iPhone 5 was the first Apple phone to come out in September, which has become a regular thing. But two years after the iPhone 5, Apple released two more models in the same line. In 2014, the iPhone 6 Plus came out simultaneously with the iPhone 6. It was advertised as "bigger than bigger" and "the two and only."

The iPhone 6 Plus set the standard for screen size, so the iPhone 7 Plus and iPhone 8 Plus also have 5.5-inch screens. The regular iPhone 6, iPhone 7, and iPhone 8 have all had 4.7-inch screens.

The iPhone 6 and iPhone 6 Plus had bigger screens, faster processors, better cameras, and better LTE and Wi-Fi connectivity. The iPhone 7 and iPhone 7 Plus came in new colors and were resistant to water and dust. The 3.5mm headphone jack was also taken away.

September 2017:

September 2017: "Say hi to the future." At Apple's first-ever event at the Steve Jobs Theater in Cupertino, California, the Apple TV 4K, Apple Watch Series 3, iPhone 8 and 8 Plus, and iPhone X were shown to the world for the first time. The iPhone X, which came out in November, was the highlight of a historic event to mark the 10th anniversary of the first iPhone.

The iPhone X was the first Apple phone with an OLED screen, a screen that went all the way to the edge, and wireless charging. Face ID lets people unlock their phones by looking at them, and Animojis move like them. The home button was taken away. Consumer Reports said that the iPhone X is the best smartphone camera ever because of Portrait Mode, Portrait Lighting, digital image stabilization, True Depth, and optical image stabilization.

September 2018

September 2018: "Welcome to the big screen." Apple made three new phones based on the iPhone X from the year before: The new iPhones are the XS, the XS Max, and the XR. The iPhone XS & XS Max have the biggest screens that any iPhone has ever had. They also have faster Face ID, a smarter and more powerful chip, and a first-of-its-kind dual-camera system. Instead, the iPhone XR has a brand-new Liquid Retina display that lets you see colors that look just like real life. It also has the largest LCD screen ever on an iPhone.

From Face Time to Face ID, the iPhone has changed how we use smartphones in a way that won't go away. And because of these new ideas and their success, the popular iPhone SE and C series were made.

October 2020

"Hello 5G" in October 2020. With the release of the iPhone 12 and the iPhone Pro Max, some new colors and four phones all support 5G. The new series has a Ceramic Shield that protects against drops 4 times better. It also has an A14 Bionic chip, a 16-core Neural Engine, and is very efficient, so it has a long battery life. 5G has superfast speeds and low latency so you can get faster downloads, better quality video streaming, more responsive gaming, and real-time interaction.

The iPhone 12 has a brand-new Ultra-Wide camera that can now take photos in Night mode and a Wide camera that can now take in 27% more light.

But even though Apple has some great phones right now, the future looks even better. If their past work is any indication, there will be many exciting technological advances soon. So, check out all the great Apple devices and accessories at Verizon, where you can get the newest phones on the country's most reliable 4G and 5G networks.

September 2021:

September 2021: Welcome, iPhone 13 and Friends. The iPhone 13 and iPhone 13 Pro came out in September 2021. They had the most advanced dual-camera system Apple had ever made for an iPhone. With a durable design, an A15 bionic chip, and a big improvement in battery life, the iPhone 13, iPhone 13 Pro, iPhone 13 Pro Max, and iPhone 13 mini are the best iPhones to date. Need to know which iPhone model to buy? Here, you can compare iPhone models and pair them with Verizon, the network that most people in the U.S. use.

1. Some places have access to 5G Ultra Wideband. More than 2700 cities have 5G all over the country.

2. Display: The display has rounded corners that are set in a standard rectangle and have a beautiful curved design. The screen's diagonal size is 6.06 inches for the iPhone 14, 6.68 inches for the iPhone 14 Plus, 6.12 inches for the iPhone 14 Pro, or 6.69 inches for the iPhone 14 Pro Max. Less can be seen.

3. SOS Satellite disclaimer: SOS via satellite for an emergency: In stores by November 2022. When you turn on any iPhone 14 model, you get two years of service for free.

Connection and response times depend on where you are, how the site is set up, and other things. Find out more at apple.com/iphone-14 or apple.com/iphone-14-pro.

iPhone (1st gen)- iPhone 3G-iPhone 3GS-iPhone 4 (GSM) iPhone 4 (CDMA)-iPhone 4S-iPhone 5-iPhone 5C-iPhone 5S-iPhone 6-iPhone 6 Plus- iPhone 6S- iPhone 6S Plus-iPhone SE (1st gen)-iPhone 7-iPhone 7 Plus-iPhone 8- iPhone 8 Plus-iPhone X-iPhone XR-iPhone XS-iPhone XS Max-iPhone 11-iPhone 11 Pro-iPhone 11 Pro Max-iPhone SE (2nd gen)-iPhone 12-iPhone 12 mini-iPhone 12 Pro-iPhone 12 Pro Max-iPhone 13-iPhone 13 mini-iPhone 13 Pro-iPhone 13 Pro Max-iPhone SE (3rd gen)-iPhone 14-iPhone 14 Plus-iPhone 14 Pro-iPhone 14 Pro Max.

iPhone (1st generation)

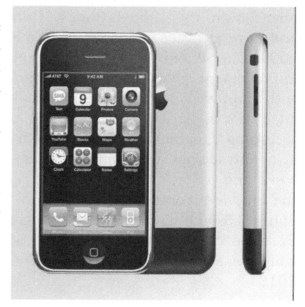

The iPhone (1st generation) is a smartphone designed and marketed by Apple Inc. It was the first iPhone released and was introduced on June 29, 2007. It was originally available in the United States, Germany, and the United Kingdom. The iPhone was the first to use a multi-touch interface and was the first to combine a phone, iPod, and internet communication device into one device. It had a 3.5-inch display and a 2-megapixel camera. It was powered by a 412 MHz ARM processor and had 128 MB of RAM. The iPhone 1st generation ran on the iOS operating system based on Mac OS X. The iPhone 3G succeeded the first generation iPhone in 2008.

The iPhone 3G

The iPhone 3G is a smartphone designed and marketed by Apple Inc. It is the second-generation iPhone and was introduced on June 9, 2008, at the Worldwide Developers Conference. It was released in the United States, Canada, and several other countries on July 11, 2008. The iPhone 3G significantly improved over the original iPhone, supporting faster 3G data speeds and assisted GPS navigation. It also introduced the App Store, which allowed users to download and install third-party applications on their iPhones. The iPhone 3G had a 3.5-inch display and a 2-megapixel camera. It was powered by a 412 MHz ARM processor and had 128 MB of RAM. It ran on the iOS operating system based on Mac OS X. The iPhone 3GS succeeded the iPhone 3G in 2009.

iPhone 3GS

The iPhone 3GS is a smartphone that was designed and marketed by Apple Inc. It is the third-generation iPhone and was introduced on June 8, 2009, at the Worldwide Developers Conference. It was released in the United States, Canada, and several other countries on June 19, 2009. The iPhone 3GS was a significant improvement over the previous models, with improved performance and features such as a higher-resolution camera, video recording capability, and voice control. It also introduced features such as cut, copy, and paste, and the ability to search the phone's contents. The iPhone 3GS had a 3.5-inch display and a 3-megapixel camera. It was powered by a 600 MHz ARM Cortex-A8 processor and had 256 MB of RAM. It ran on the iOS operating system, which was based on Mac OS X. The iPhone 3GS was succeeded by the iPhone 4 in 2010.

iPhone 4 (GSM)

The iPhone 4 (GSM) is a smartphone that was designed and marketed by Apple Inc. It is the fourth-generation iPhone and was introduced on June 7, 2010, at the Worldwide Developers Conference. It was released in the United States, Canada, and several other countries on June 24, 2010. The iPhone 4 was a significant improvement over previous models, with a new design, higher-resolution display, and improved performance. It also introduced the front-facing camera, which allowed for video calls, and the Retina display, which had a higher pixel density than previous models. The iPhone 4 (GSM) had a 3.5-inch display and a 5-megapixel camera. It was powered by an Apple A4 processor and had 512 MB of RAM. It ran on the iOS operating system, which was based on Mac OS X. The iPhone 4 (GSM) was succeeded by the iPhone 4S in 2011.

iPhone 4 (CDMA)

The iPhone 4 (CDMA) is a smartphone designed and marketed by Apple Inc. It is similar to the iPhone 4 (GSM) but was designed for CDMA networks rather than GSM networks. It was released on February 10, 2011, and was available in several other countries. The iPhone 4 (CDMA) had a 3.5-inch display and a 5-megapixel camera. It was powered by an Apple A4 processor and had 512 MB of RAM. It ran on the iOS operating system based on Mac OS X. The iPhone 4 (CDMA) was succeeded by the iPhone 4S in 2011.

iPhone 4S

The iPhone 4S was a smartphone that was released in 2011 by Apple. It was the successor to the iPhone 4 and was the last iPhone model to use the 30-pin connector. It had a 3.5-inch Retina display, a dual-core A5 processor, and an 8-megapixel camera. It was available with 16GB, 32GB, or 64GB of storage. The iPhone 4S also introduced Siri, a virtual assistant that could be activated with a voice command.

iPhone 5

The iPhone 5 was a smartphone that was released in 2012 by Apple. It was the successor to the iPhone 4S and was the first iPhone to be released with a larger 4-inch display. It had a dual-core A6 processor and an 8-megapixel camera. It was available with 16GB, 32GB, or 64GB of storage. The iPhone 5 was also the first to be released with a lightning connector, which replaced the 30-pin connector used on previous models.

iPhone 5C

The iPhone 5C was a smartphone that was released by Apple in 2013. It was similar to the iPhone 5, but it featured a plastic back panel and was available in a range of colors. It had a 4-inch Retina display, a dual-core A6 processor, and an 8-megapixel camera. It was available with either 16GB or 32GB of storage. The iPhone 5C was intended to be a more affordable alternative to the iPhone 5S, which was released at the same time.

iPhone 6

The iPhone 6 was a smartphone that was released by Apple in 2014. It had a 4.7-inch Retina HD display, a dual-core A8 processor, and an 8-megapixel camera. It was available with 16GB, 64GB, or 128GB of storage. The iPhone 6 also introduced Apple Pay, a mobile payment service that allowed users to make payments

3

with their iPhone. It was the first iPhone to be released with a larger screen size since the iPhone 5 and was also available in a larger "Plus" size with a 5.5-inch display.

iPhone 6 Plus

The iPhone 6 Plus was a smartphone that was released by Apple in 2014. It was similar to the iPhone 6, but it had a larger 5.5-inch Retina HD display and a longer battery life. It had a dual-core A8 processor and an 8-megapixel camera. It was available with 16GB, 64GB, or 128GB of storage. The iPhone 6 Plus was the first iPhone to be released with a phablet-sized display and was also the first iPhone to offer optical image stabilization for its camera. It was intended to be a larger and more powerful alternative to the smaller iPhone 6.

iPhone 6S

The iPhone 6S was a smartphone that was released by Apple in 2015. It was an updated version of the iPhone 6, featuring improved hardware and software. It had a 4.7-inch Retina HD display, a dual-core A9 processor, and a 12-megapixel camera. It was available with 16GB, 64GB, or 128GB of storage. The iPhone 6S also introduced 3D Touch, a pressure-sensitive screen technology that enabled new types of gestures and interactions. It was available in the same sizes and configurations as the iPhone 6.

iPhone 6S Plus

The iPhone 6S Plus was a smartphone that was released by Apple in 2015. It was similar to the iPhone 6S, but it had a larger 5.5-inch Retina HD display and a longer battery life. It had a dual-core A9 processor and a 12-megapixel camera. It was available with 16GB, 64GB, or 128GB of storage. The iPhone 6S Plus also featured 3D Touch, a

pressure-sensitive screen technology that enabled new types of gestures and interactions. It was intended to be a larger and more powerful alternative to the smaller iPhone 6S.

iPhone SE (1ˢᵗ generation)

The first-generation iPhone SE was a smartphone that was released by Apple in 2016. It was a budget-friendly alternative to Apple's more expensive iPhone models, featuring many of the same internal components as the iPhone 6S but in a smaller, more compact body similar to the iPhone 5S. It had a 4-inch Retina display, a dual-core A9 processor, and a 12-megapixel camera. It was available with either 16GB or 64GB of storage. The iPhone SE was popular with users who preferred the compact size and design of the iPhone 5S but wanted the more powerful hardware of the iPhone 6S.

iPhone 7

The iPhone 7 was a smartphone that was released by Apple in 2016. It had a 4.7-inch Retina HD display, a quad-core A10 Fusion processor, and a 12-megapixel camera. It was available with 32GB, 128GB, or 256GB of storage. The iPhone 7 was the 1st iPhone to be released without a traditional headphone jack, instead requiring users to use the Lightning port or a separate adapter to connect their headphones. It was also the 1st iPhone to be water-resistant. The iPhone 7 was succeeded by the iPhone 8 in 2017.

iPhone 7 Plus

The iPhone 7 Plus was a smartphone that was released by Apple in 2016. It was similar to the iPhone 7, but it had a larger 5.5-inch Retina HD display and a dual-camera system. It had a quad-core A10 Fusion processor and was available with 32GB, 128GB, or 256GB of storage. The dual-camera system on the iPhone 7 Plus consisted of a wide-angle lens and a telephoto lens, which allowed for optical zoom at 2x and up to 10x digital zoom. Like the iPhone 7, the iPhone 7 Plus was the first iPhone to be released without a traditional headphone jack and was also water-resistant. The iPhone 7 Plus was succeeded by the iPhone 8 Plus in 2017.

iPhone 8

The iPhone 8 was a smartphone that was released by Apple in 2017. It had a 4.7-inch Retina HD display and was powered by a hexa-core A11 Bionic processor. It had a single 12-megapixel camera and was available with either 64GB or 256GB of storage. The iPhone 8 featured a glass back panel, which allowed for wireless charging. It was the first iPhone to support wireless charging and also the first iPhone to be powered by the A11 Bionic chip. The iPhone 8 was succeeded by the iPhone X in 2017.

iPhone 8 Plus

The iPhone 8 Plus was a smartphone that was released by Apple in 2017. It was similar to the iPhone 8, but it had a larger 5.5-inch Retina HD display and a dual-camera system. It was powered by a hexa-core A11 Bionic processor and was available with either 64GB or 256GB of storage. The dual-camera system on the iPhone 8 Plus consisted of a wide-angle lens and a telephoto lens, which allowed for optical zoom at 2x and up to 10x digital zoom. The iPhone 8 Plus also featured a glass back panel, which allowed for wireless charging. It was the first iPhone to support wireless charging and was also powered by the A11 Bionic chip. The iPhone 8 Plus was succeeded by the iPhone X in 2017.

iPhone X

The iPhone X (pronounced "ten") was a smartphone that was released by Apple in 2017. It was a significant departure from previous iPhone models, featuring a full-screen display with a "notch" at the top for front-facing cameras and sensors. It had a

5.8-inch Super Retina OLED display and was powered by a hexa-core A11 Bionic processor. It had a dual-camera system consisting of a wide-angle lens and a telephoto lens and was available with either 64GB or 256GB of storage. Face ID, a facial recognition system that could be used to unlock the phone and pay for things, was also added to the iPhone X. It was the first iPhone to use a full-screen display and to have Face ID. The iPhone X was succeeded by the iPhone XS in 2018.

iPhone XR

The iPhone XR was a smartphone that was released by Apple in 2018. It was a lower-cost alternative to the iPhone XS and featured many of the same internal components, including the A12 Bionic processor and the TrueDepth camera system for Face ID. It had a 6.1-inch Liquid Retina LCD display and a single 12-megapixel camera. It was available with 64GB, 128GB, or 256GB of storage. The iPhone XR was available in a range of colors and was the first iPhone to be offered in red as part of the Product RED program. It was also the 1st iPhone to be offered in blue, coral, and yellow.

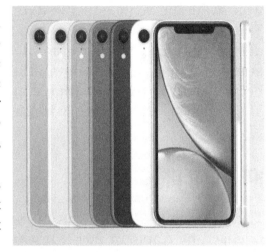

iPhone XS

The iPhone XS is a smartphone designed and developed by Apple Inc. It was announced on September 12, 2018, alongside the iPhone XS Max and the iPhone XR. The iPhone XS features a 5.8-inch Super Retina OLED display, with a resolution of 2436 x 1125 pixels. It is powered by a 7nm A12 Bionic chip with a 6-core CPU and a 4-core GPU and has 4GB of RAM. The phone also has a 12-megapixel dual-camera system on the back, and there is a 7-megapixel TrueDepth camera on the front for Face ID and Animoji. The iPhone XS runs on Apple's latest operating system, iOS 12.

iPhone XS Max

The iPhone XS Max is a smartphone designed and developed by Apple Inc. It was announced on September 12, 2018, alongside the iPhone XS and the iPhone XR. The iPhone XS Max features a 6.5-inch Super Retina OLED display, the largest display ever on an iPhone, with a resolution of 2688 x 1242 pixels. It is powered by a 7nm A12 Bionic chip with a 6-core CPU and a 4-core GPU and has 4GB of RAM. The phone also has a 12-megapixel dual-camera system on the back, and there is a 7-megapixel TrueDepth camera on the front for Face ID and Animoji. The iPhone XS Max runs on Apple's latest operating system, iOS 12.

iPhone 11

The iPhone 11 is a smartphone designed and developed by Apple Inc. It was announced on September 10, 2019, alongside the iPhone 11 Pro and the iPhone 11 Pro Max. The iPhone 11 features a 6.1-inch Liquid Retina LCD display, with a resolution of 1792 x 828 pixels. It is powered by a 7nm A13 Bionic chip with a 6-core CPU and a 4-core GPU and has 4GB of RAM. The phone also has a 12-megapixel dual-camera system on the back, consisting of a primary camera and an ultra-wide-angle camera, and there is a 12-megapixel TrueDepth camera on the front for Face ID and Animoji. The iPhone 11 runs on Apple's latest operating system, iOS 13.

iPhone 11 Pro

The iPhone 11 Pro is a smartphone designed and developed by Apple Inc. It was announced on September 10, 2019, alongside the iPhone 11 and the iPhone 11 Pro Max. The iPhone 11 Pro has a Super Retina XDR OLED display that measures 5.8 inches and has a resolution of 2436 x 1125 pixels. It is powered by a 7nm A13 Bionic chip with a 6-core CPU and a 4-core GPU and has 4GB of RAM. The phone also has a triple-camera system on the back, consisting of a primary camera, an ultra-wide-angle camera, and a telephoto camera, and there is a 12-megapixel TrueDepth camera on the front for Face ID and Animoji. The iPhone 11 Pro runs on Apple's latest operating system, iOS 13.

iPhone 11 Pro Max

iPhone 11 Pro Max is a smartphone designed and developed by Apple Inc. It was announced on September 10, 2019, alongside the iPhone 11 and the iPhone 11 Pro. The iPhone 11 Pro Max features a 6.5-inch Super Retina XDR OLED display, the largest display ever on an iPhone, with a resolution of 2688 x 1242 pixels. It is powered by a 7nm A13 Bionic chip with a 6-core CPU and a 4-core GPU and has 4GB of RAM. The phone also has a triple-camera system on the back, consisting of a primary camera, an ultra-wide-angle camera, and a telephoto camera, and there is a 12-megapixel TrueDepth camera on the front for Face ID and Animoji. The iPhone 11 Pro Max runs on Apple's latest operating system, iOS 13.

iPhone SE (2nd generation)

The iPhone SE (2nd generation) is a smartphone designed and developed by Apple Inc. It was announced on April 15, 2020, as a successor to the original iPhone SE, which was released in 2016. The 2nd generation iPhone SE features a 4.7-inch Retina HD LCD with a resolution of 1334 x 750 pixels. It is powered by a 7nm A13 Bionic chip with a 6-core CPU and a 4-core GPU and has 3GB of RAM. The phone also has a 12-megapixel camera on the back and a 7-megapixel TrueDepth camera on the front for Face ID and Animoji. The iPhone SE (2nd generation) runs on Apple's latest operating system, iOS 13.

iPhone 12

The iPhone 12 is a smartphone that was made by Apple Inc. It was announced on October 13, 2020, along with the iPhone 12 Mini, the iPhone 12 Pro, and the iPhone 12 Pro Max. The iPhone 12 features a 6.1-inch Super Retina XDR OLED display with 2532 x 1170 pixels. It is powered by a 5nm A14 Bionic chip with a 6-core CPU and a 4-core GPU and has 4GB of RAM. The phone also has a dual-camera system on the back, consisting of a primary camera and an ultra-wide-angle camera, and there is a 12-megapixel TrueDepth camera on the front for Face ID and Animoji. The iPhone 12 runs on Apple's latest operating system, iOS 14.

iPhone 12 Mini

The iPhone 12 Mini is a smartphone designed and developed by Apple Inc. It was announced on October 13, 2020, alongside the iPhone 12, the iPhone 12 Pro, and the iPhone 12 Pro Max. The iPhone 12 Mini features a 5.4-inch Super Retina XDR OLED display with a 2340 x 1080 pixels resolution. It is powered by a 5nm A14 Bionic chip with a 6-core CPU and a 4-core GPU and has 4GB of RAM. The phone also has a dual-camera system on the back, consisting of a primary camera and an ultra-wide-angle camera, and there is a 12-megapixel TrueDepth camera on the front for Face ID and Animoji. The iPhone 12 Mini runs on Apple's latest operating system, iOS 14. c-iPhone 12 Pro Max-iPhone 13-iPhone 13 mini-iPhone 13 Pro-iPhone 13 Pro Max-iPhone SE (3rd gen)-iPhone 14-iPhone 14 Plus-iPhone 14 Pro-iPhone 14 Pro Max

iPhone 12 Pro

The iPhone 12 Pro is a smartphone that Apple Inc made. On October 13, 2020, it was announced along with the iPhone 12, iPhone 12 Mini, and iPhone 12 Pro Max. The iPhone 12 Pro features a 6.1-inch Super Retina XDR OLED display with 2532 x 1170 pixels. It is powered by a 5nm A14 Bionic chip with a 6-core CPU and a 4-core GPU and has 6GB of RAM. The phone also has a triple-camera system on the back, consisting of a primary camera, an ultra-wide-angle camera, a telephoto camera, and a 12-megapixel TrueDepth camera on the front for Face ID and Animoji. The iPhone 12 Pro runs on Apple's latest operating system, iOS 14.

iPhone 12 Pro Max

Apple Inc. has created a new smartphone called the iPhone 12 Pro Max. As well as the iPhone 12, iPhone 12 Mini, and iPhone 12 Pro, it was launched on October 13, 2020. The iPhone 12 Pro Max features a 6.7-inch Super Retina XDR OLED display, the largest display ever on an iPhone, with a resolution of 2778 x 1284 pixels. It is powered by a 5nm A14 Bionic chip with a 6-core CPU and a 4-core GPU and has 6GB of RAM. The phone also has a triple-camera system on the back, consisting of a primary camera, an ultra-wide-angle camera, and a telephoto camera, and there is a 12-megapixel TrueDepth camera on the front for Face ID and Animoji. The iPhone 12 Pro Max runs on Apple's latest operating system, iOS 14.1

iPhone 13 series

On September 14, 2021, at a California Streaming event, Apple introduced the iPhone 13 series. Since Apple unveiled a new design for its high-end iPhones last year with the iPhone 12 series and often sticks with the same design for up to three years in a row, the company did not make any changes to the design this year. New and improved hardware has been offered by Apple, though. Apple introduced not one, not two, but four new iPhones this year: the iPhone 13 Mini, iPhone 13, iPhone 13 Pro, and iPhone 13 Pro Max. Check out our hands-on video of the iPhone 13 Pro and our in-depth review of the iPhone 13 Pro Max, as well as read all about them in our article down below!

A new Apple A15 Bionic engine, a 120Hz high-refresh-rate OLED display with a slightly smaller notch, upgraded cameras, and larger batteries are the most notable additions. It also included a new depth-of-field camera mode called Cinematic Video. Some of the iPhone 13 models also include quicker charging, another new feature from the firm.

IPHONE 14 AND IPHONE 14 PLUS

The 6.1 inches and 6.7-inch iPhone 14 and iPhone 14 Plus, which Apple unveiled, come in two sizes and have a clever design, spectacular camera upgrades, and revolutionary new safety features. The iPhone 14 and iPhone 14 Plus have a powerful camera system that incorporates the Main and front True Depth cameras, the Ultra-Wide camera for unique vistas, and Photogenic Engine—an upgraded picture pipeline—that produces stunning photos and movies. The A15 Bionic CPU, which includes a 5-core GPU in every model, is designed with privacy and security in mind and provides exceptional speed and efficiency for demanding applications. The iPhone 14 and iPhone 14 Plus are the first devices in their class to integrate critical safety features like Crash Detection and Emergency SOS via satellite. The exceptional battery life, industry-leading durability features, and lightning-fast 5G make this iPhone series more advanced than ever. Customers depend on their iPhones every day, and the iPhone 14 and iPhone 14 Plus offer revolutionary new technology and essential safety features. According to Greg Joswiak, senior vice president of worldwide marketing at Apple, users can view more web content and text on the larger 6.7-inch display on the iPhone 14 Plus. Both phones include a powerful new Main camera that performs far better in low light, improved connectivity choices with 5G and e SIM, and the incredible A15 Bionic performance, which helps to extend battery life even further. The intimate connection between iOS 16 and the make of iPhone is more important than ever.

A Gorgeous and Sturdy Design with Incredible Battery Life

The iPhone 14 and iPhone 14 Plus come in well-liked 6.1-inch and magnificent new 6.7-inch sizes2 and have a sturdy and svelte aerospace-grade aluminum body in five lovely colors. The iPhone 14 Plus has the best battery life of any iPhone ever, and its larger display is fantastic for watching movies and playing games. In addition to stunning Super Retina XDR displays with OLED technology that supports Dolby Vision, a contrast ratio of 2,000,000:1, 1200 nits of peak HDR brightness, and stunning Super Retina XDR displays with OLED technology, both models have improved internal architecture for better thermal performance. Upgraded Cameras with Powerful Photogenic Engine

With a new 12MP main camera that has a larger sensor and larger pixels, a new front True Depth camera, the Ultra-Wide camera to capture more of a scene, and the Photogenic Engine for a significant improvement in low-light performance, the iPhone 14 and iPhone 14 Plus set a new standard for photo and video capture.

Upgrades and improvements for dual-camera systems include:
- The Ultra-Wide camera
- The new Action mode
- The new Main camera
- The front True Depth camera
- The upgraded True Tone flash
- The Cinematic mod.

Emergency SOS and Crash Detection through Satellite

Groundbreaking safety features that can assist in an emergency when it matters the most are available across the entire iPhone 14 family. When a user is unconscious or unable to reach their iPhone, Crash Detection on the iPhone may instantly notify emergency services and identify a serious auto accident thanks to a new high dynamic range gyroscope and dual-core accelerometer that can detect G-force values of up to 256Gs. The barometer, which can now detect variations in cabin pressure; the GPS, which offers additional input for speed changes; and the microphone, which can recognize

loud noises typical of catastrophic auto accidents, are examples of functions that these capabilities have already enhanced.

Strong Connection Features

To let consumers keep in touch, share material, and enjoy it, iPhone gives users incredibly fast download and upload speeds, improved streaming, and real-time connectivity with 5G. With extended support for independent networks, 5G support for the iPhone is currently available from more than 250 carrier partners in more than 70 regions worldwide. With e-SIM, users may rapidly connect to or transfer their current plans digitally with support for multiple cellular plans on a single device. It is a more secure alternative to a physical SIM card. With the SIM tray gone on the iPhone 14 and iPhone 14 Plus for US models, users may quickly and easily set up their devices.

All iPhone Users Can Use Apple Fitness

Later this fall, Apple Fitness+ will be made available to all iPhone users in the 21 nations where it is currently available, regardless of whether they have an Apple Watch. iPhone users will get access to the entire program, which includes more than 3,000 studio-style exercises and meditations led by a diverse and inclusive roster of trainers. Onscreen trainer instructions and interval timing will be provided to Fitness+ users, and progress on their Move ring will be monitored using estimated calories burned. Prizes, activity sharing, and other features in the center tab of the Fitness+ app, which will be completely integrated with the Fitness app that launches with iOS 16, can help users stay motivated to close their Move ring.

APPLE IPHONE 14 AND IPHONE PRO MAX

Apple today revealed the iPhone 14 Pro and iPhone 14 Pro Max, two of the most technologically advanced Pro models yet. These models include Dynamic Island and Always-On displays. The first-ever 48MP main camera on an iPhone with a quad-pixel sensor and Photogenic Engine, an upgraded picture pipeline that greatly improves low-light photos, debuts in the iPhone 14 Pro, which is powered by the A16 Bionic, the fastest chip ever in a smartphone. With features like Emergency SOS via satellite and Crash Detection, iPhone has expanded its usefulness beyond normal tasks to include emergency situations. Customers depend on their iPhone every day, and we're introducing more innovations to the market with the iPhone 14 Pro and iPhone 14 Pro Max than ever before. "iPhone 14 Pro introduces a camera system that empowers every user, from the casual user to the professional, to take their best photos and videos, as well as cutting-edge new technologies like the Always-On display and the Dynamic Island, which offers new interactions for notifications and activities," said Greg Joswiak, senior vice president of worldwide marketing at Apple. Innovative safety features increase people's sense of security and offer help when they most need it. And because of the A16 Bionic chip's superior power and efficiency as well as its all-day battery life, this iPhone is the best one yet.

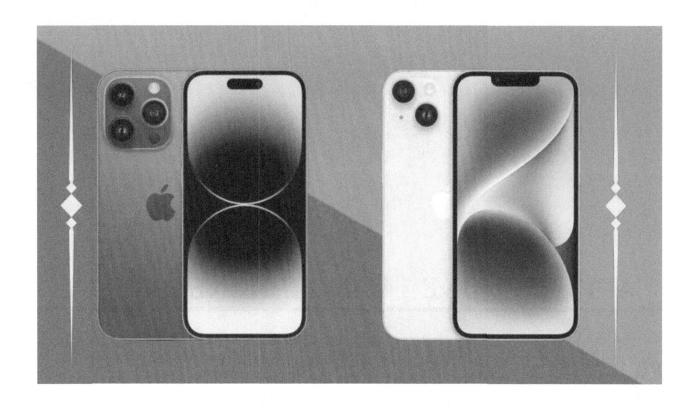

The Most Modern Smartphone Display and a Stunning Design

The iPhone 14 Pro and iPhone 14 Pro Max have stunning designs made of surgical-grade stainless steel and textured matte glass that are available in four magnificent colors. The new Super Retina XDR display with Pro Motion found in both models—available in 6.1-inch and 6.7-inch sizes1—offers the Always-On display for the first time ever on an iPhone thanks to a new 1Hz refresh rate and other power-saving advancements. The redesigned Lock Screen is much more practical thanks to the constant availability of time, widgets, and Live Activities. As bright as 2000 nits, which is twice as bright as the iPhone 13 Pro and has the same peak HDR brightness level as Pro Display XDR, the cutting-edge display also boasts the highest outdoor peak brightness in a smartphone.

Big Changes

The iPhone 14 Pro and iPhone 14 Pro Max significantly alter Apple's phone design in a number of ways.

✪ Photos may be taken in Pro Raw mode on the phones at their full 48-megapixel resolution.

- The phones contain a 48-megapixel primary camera on the back as well as 12-megapixel ultrawide and 12-megapixel telephoto lenses.

- Additionally, both iPhone 14 Pro phones support 4K, 24 fps Cinematic style video recording.

- The new A16 Bionic chip from Apple is included in the iPhone 14 Pro and Pro Max handsets.

- Enhancing color and detail in images is also possible with Apple's brand-new Photogenic Engine image processing.

- Coming around to the front reveals the new screen cutout on the Pro and Pro Max, where the Dynamic Island feature is located.

- As you switch between apps, the Dynamic Island displays pertinent information. When you start a song in Apple Music, for instance, it will display the music that is now playing.

- Depending on what may be displayed, the Dynamic Island will appear to shrink and grow along the cutout.

- The Pro models have satellite connectivity and car crash detection, just like the iPhones 14 and 14 Plus.

- iOS 16 will come preinstalled on Apple's iPhone 14 Pro and Pro Max handsets.

EASIEST WAY TO SET UP YOUR NEW IPHONE 14

What to do prior to configuring your iPhone:

Before starting, you'll need the following items if you're moving from another phone or setting up an iPhone for the first time:

Back up your old iPhone - It is a good idea to have your old phone ready, whether it's an Android or an iPhone. In fact, doing so might speed up the setup procedure.

Keep your old phone close by- Having your old phone available is a smart idea whether you're switching to an iPhone or an Android. The setup procedure may even be sped up by doing this.

And have your charger with you, just in case-You should have enough battery on your new iPhone 14 to get you through the setup procedure, but you may need electricity to charge your older iPhone while making the transfer.

Make sure you are connected to the Internet- To configure your iPhone, you must connect to a reliable Wi-Fi network.

Have your Apple ID credentials on hand-Have your email address and password handy in case you need to log into your Apple account to restore a backup or do anything else. One can be made even more easily during setup.

1. Turn on your iPhone 14

Turning on your iPhone is the first thing you must do. Regardless of the model, turning it on is the same: When you see the Apple logo, press and hold the side button on the right side of your iPhone. The Hello screen should show after a short while. Up swipe to start.

2. Follow the iPhone onscreen instructions

To complete the first step of the setup, you will now need to do a handful of quick and simple tasks on your iPhone. You will first need to

Pick a language: If you're in the US, English should be listed first and center.

Choose your nation or region. Again, if you are in the US, "United States" should be visible at the top.

3. Use Quick Start to set up your iPhone 14

The Quick Start screen, which lets you automatically set up your iPhone 14 with the aid of your old iPhone or even an iPad, can be found on the following page. This will require you to turn on your old smartphone, connect it to Wi-Fi or cellular, and then bring it close to the new iPhone 14. The two phones must then be connected or authenticated after that. Either enter a verification code or use the camera on your old iPhone to scan the pattern on the new iPhone. Then, it will take some time for activation. Following that, you'll be asked to take the following actions, among others:

On your new iPhone, connect to your Wi-Fi network.

Install eSIM. Two choices ought to appear: Transfer from a different iPhone is the best choice, followed by Setting Up Later in Settings. You'll be prompted to double-press the side button on your old iPhone if you select option one.

Install Touch ID or Face ID.

Select the data transfer method of your choice. You have the option of doing it from your old iPhone or from iCloud.

It is a good idea to keep both smartphones powered on while you do this because it can take some time depending on how many apps and how much data you have.

4. Set your iPhone 14 up manually

Now, on the Quick Start screen, select Set Up Manually if you're switching from an Android or another phone or wish to manually configure your iPhone. If you choose this course of action, you will need to manually complete the following tasks

Configure a Wi-Fi network. The activation of your phone may take a while.

Read the prompt for Data & Privacy. When you're done, click Continue.

Set up Touch ID or Face ID. You can use your fingerprint or facial recognition to unlock your iPhone 14 in this way.

To unlock your iPhone 14, create a six-digit pass code. Your fallback method for unlocking your iPhone 14 is this. Enter the pass code twice.

Pick a recovery method for your apps and data. You can transfer directly from your iPhone, move data from your Android device, restore from an iCloud backup, restore from a Mac or PC, or choose not to transfer any applications or data.

You might be prompted for your iCloud login information to access your Apple account, depending on how you choose to restore your applications and data. To confirm your identification, you must then enter a verification code that can appear on one of your other devices.

5. Now use your iPhone 14 running iOS 16

Once you are finished, to enter your iPhone, swipe up. You'll need to wait while your phone gradually uploads all of your apps and data to your new iPhone 14. You might not see all of your program on your home screen or all of your images and videos in your camera roll. Depending on how much needs to be sent, this could take several hours.

HOW TO USE SIRI ON IPHONE 14

You can use Siri to plan meetings, make phone calls, send text and audio messages, and more. Siri is a smart personal assistant that you can use to accomplish tasks by asking. You may use your voice to set up appointments, make phone calls, and more. Siri, however, differs from conventional voice-recognition software in that it doesn't require you to speak particular orders or recall a list of terms. Siri can comprehend your natural speech; if it requires more details to do a task, it will enquire about you. It can intelligently combine your regular activities with outside apps to provide practical shortcuts when you need them.

Use Siri

1. To ask Siri a question, hold the Side button until the Siri icon appears at the bottom of the screen.

You must activate Siri first. To do this, open the Settings app, scroll to Siri & Search, and then pick the off option. To switch to Siri, press the side button. To confirm, select Enable Siri. Activate the switch. Siri may be turned on or off from the lock screen using the Allow Siri When Locked switch.

2. If you're using iOS 8 or later, you don't need to hold down the Side button to ask Siri a question that begins with "Hey Siri."

HOW TO CLOSE OR CLEAR AN APP ON IPHONE 14

On your iPhone 14, you must utilize specific swipe movements to view open apps, end running apps, and exit of program. The process is precisely the same for older iPhone models, including the iPhone 11, iPhone 12, and iPhone 13. This short tutorial on the iPhones 14, 14 Plus, 14 Pro, or 14 Pro Max explains how to force close an app, exit an app, and view and close background apps.

How to exit an app on iPhone 14

Swipe up from the bottom edge of your iPhone's screen to exit an app and return to the home screen. While the specific program you were using will be closed, as a result, it might still be running in the background.

How to force close apps on iPhone 14

An app may occasionally stop working or become stuck on the loading or waiting screen. If this occurs, you may force close the program and reopen it using the App Switcher.

Users may kill non-responding programs and end background apps by using a force stop. It is helpful when you need to unfreeze an iPhone app.

- ✪ Press and hold the Home button on your iPhone 14 or iPhone 14 Pro.
- ✪ Swipe upward from the bottom of the screen then stop at the center.
- ✪ To find the app you want to force close, swipe right or left.
- ✪ Swipe up on the app's preview to force an app to close.

How to close apps on iPhone 14 all at once

On iPhone 14, you can close background programs, but iOS does not provide a means to close all open apps on an iPad or iPhone. Therefore, with iPhone 14 or any other iPhone, you cannot close all apps at once. However, gesture-based navigation allows you to forcefully close one or more apps simultaneously.

On the iPhone 14 or 14 Pro, swipe up from the bottom and pause in the screen's middle to close multiple apps. All open apps are now shown in the App Switcher. Put three fingers on each app preview simultaneously, and swipe up to close each app individually.

HOW TO FORCE RESTART IPHONE 14

The following describes how the iPhone 14 Pro, iPhone 14 Pro Max, iPhone 14, and iPhone 14 Plus handle forced restarts:

- ✪ Release the Volume Up button after pressing it.
- ✪ Release the Volume Down button after pressing it.
- ✪ Hold the power/Lock button down while pressing and holding it until the Apple logo appears on the screen.
- ✪ You'll know the forced restart was successful when the Apple logo shows on the screen, at which point you can release all button pressure.

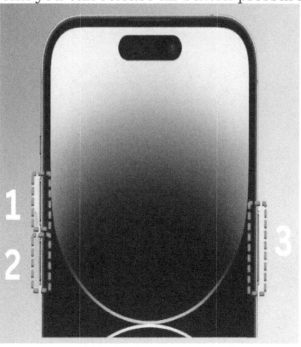

That's all there is to it; you can now restart an iPhone 14 Pro, iPhone 14 Pro Max, iPhone 14, or iPhone 14 Plus. The following describes how the iPhone 14 Pro, iPhone 14 Pro Max, iPhone 14, and iPhone 14 Plus handle forced restarts: And yes, a force restart is the same as a force reboot, and it's also commonly referred to as a force reset. However, it's important to note that nothing is truly reset on the iPhone; instead, the device abruptly goes off and back on, disrupting whatever is happening. As a typical troubleshooting technique, forcing a restart is highly useful to know how to do, so practice this a few times to learn the button combinations. The commands to remember are Volume Up, Down, and Hold Power/Lock until you see the Apple logo.

HOW TO ENABLE AND ACTIVATE I MESSAGE ON IPHONE 14

Activate and enable I Message from the Settings app.

You can enable I Message via the Settings app if it was either not enabled in some way during the setup of your iPhone 14 or you have previously turned it off. Enabling and activating I Message via the Settings app is fairly simple and won't involve any work from your end or technological know-how. Start with the Settings app from your device's home screen or store.

To proceed, select 'Messages' from the list by tapping it.

- ✪ Toggle the switch next to the "i Message" option to the "On" position by tapping it now.

- ✪ After turning it on, select 'Send & Receive' to proceed.

- ✪ To get I Messages on your device, tap the mobile number or email address given on the screen. Your contacts will see the same address displayed.

- ✪ If you have numerous addresses, choose the one you want to use when starting new chats. You can read and respond to Messages sent to any address.

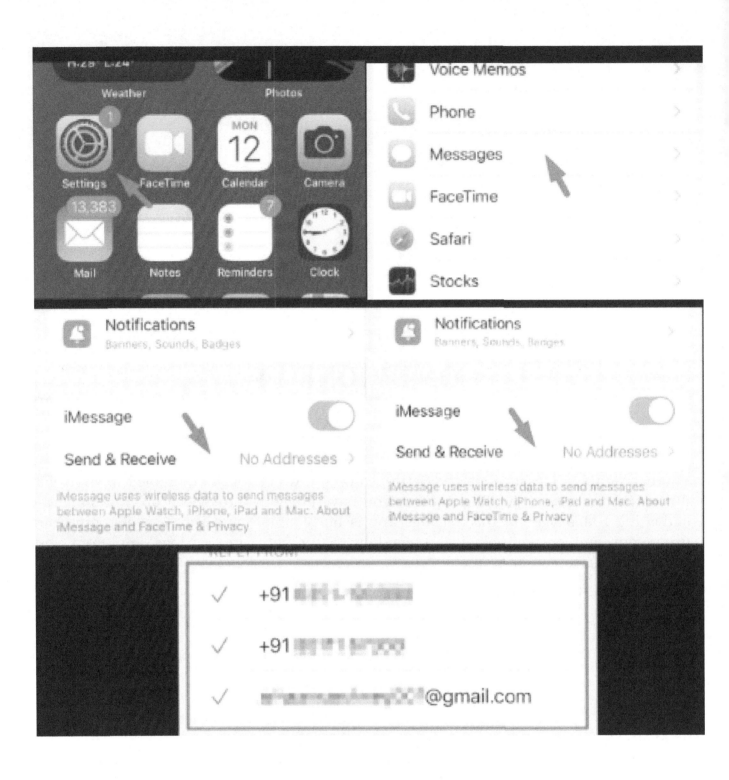

TAKING A SCREENSHOT ON AN IPHONE 14

Frequently updated screenshots and a new copy/delete option

To take a screenshot, press the side button and the volume up button simultaneously (you can also press one immediately after the other)

You can select more possibilities for five seconds by tapping the screenshot thumbnail in the bottom left corner. - If you don't do anything, the Photos app will store it.

You may mark up a screenshot by tapping the thumbnail. For quick alternatives, hit the share icon in the top right corner (a square with an up arrow).

For the convenient new Copy and Delete option, tap Done in the top left corner.

Scrolling screenshot on iPhone 14

With this native iOS feature, taking a scrolling screenshot or a lengthy snapshot will provide you a PDF of every document, web page, note, or email in one of Apple's apps (or a selected few).

- ✪ Take a picture (press the side button and volume up button at the same time)
- ✪ Before it vanishes, tap the thumbnail in the lower left corner.
- ✪ In the top right corner, select Full Page.
- ✪ The Full Page option is available only when there is more content on the screen than one page.
- ✪ Before saving, you can utilize markup, inspect every page, and more.

You can also select the portion of the Full Page you want to save by tapping the crop button (a square-shaped icon next to "Done").

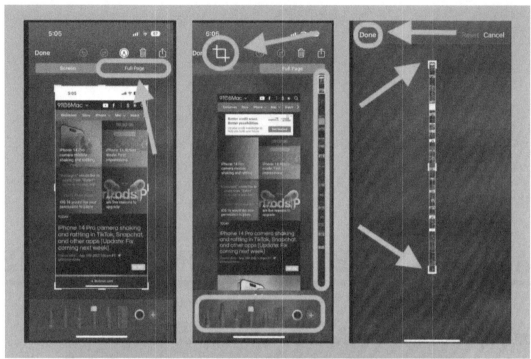

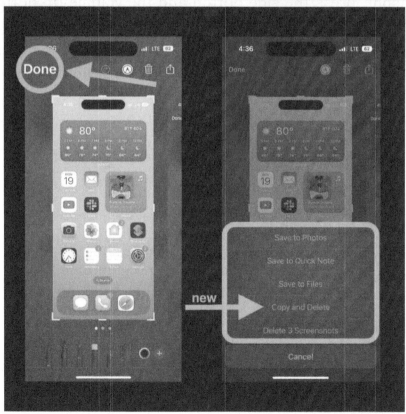

WHAT IS DYNAMIC ISLAND?

Early rumors said that Apple was working on a notch replacement with a pill-shaped cutout and a hole punch cutout to house the True Depth camera technology before the iPhone 14 Pro was formally introduced. It was in late August, a week after Apple's iPhone event invites were put out, that we made the exclusive claim that these cutouts would combine into a single, longer pill shape while the iPhone 14 Pro was in use. Additionally, we should have mentioned Apple's plans to combine software features with the pill. The iPhone 14 Pro models currently have display pixels surrounding what Apple refers to as the "Dynamic Island," which combines into a single pill-shaped area that changes in size and shape to accommodate various alerts, notifications, and interactions, turning it into a sort of front-and-center information hub.

What Can Dynamic Island Do?

Here is a breakdown of the many functions Dynamic Island can fulfill.

1. Confirmations of Apple Pay transactions
2. While a microphone or camera is being used, privacy indicators
3. Transmit files via AirDrop
4. Battery life and connection status for Air Pods
5. Low battery warnings and the state of iPhone charging
6. Whether silent mode is active or not
7. Unlocking with Face ID
8. Locking/unlocking a car key

9. Unlocking NFC connections with Apple Watch

10.   Connections via AirPlay

11.   Focusing mode shifts

12.   Quick actions

13.   Alerts for airplane mode/no data

14.   SIM card warnings

15.   Devices that link

16.   Locate My alerts

How does dynamic island work?

On Dynamic Island, some of the displayed content can be interacted with. For example, if it's showing information on an app's background activities, pressing Dynamic Island will launch that app immediately. For instance, when the media is playing, you can long-press Dynamic Island to show a widget with playback controls.

The Dynamic Island can also display multiple background events, such as when music is playing while a timer is about to expire. Due to the island's division between a larger pill-shaped portion and a smaller circular area, you can participate in both activities. Like the standard Dynamic Island interface, you can tap into and flip between them.

Which iPhone Models Feature Dynamic Island?

The only smartphones that work with Dynamic Island are the iPhone 14 Pro and the iPhone 14 Pro Max. The standard iPhone 14 units still have the same notch as the iPhone 13 models.

CAN IPHONE 14 CONNECT TO A SATELLITE?

The iPhone 14 series' four variants all come with basic satellite connectivity. Since last year, there have been reports that Apple has been working on adding a satellite connectivity capability to the iPhone. The feature is now available with the release of the iPhone 14 series. There are several restrictions and caveats in this. But Apple has managed to pull off the remarkable accomplishment of becoming the first to provide such a capability on a smartphone.

Very little at this time. The satellite connectivity on the iPhone 14 is a safety feature that enables you to contact emergency services if you are lost or stuck without cellular or internet connectivity. In these circumstances, your iPhone 14 can send a message to emergency personnel through a satellite connection. This function is known as Emergency SOS via satellite by Apple.

How does Emergency SOS via satellite work?
To use Emergency SOS via satellite, it is necessary to be in an open area with a clear view of the sky and horizon. To connect to a satellite, the iPhone 14 will provide instructions onscreen. After connecting, you'll be prompted to describe your current situation by responding to emergency questions. Your iPhone then initiates a conversation with emergency personnel by providing them with your medical ID, emergency contact information, location, altitude, remaining battery life, and emergency questionnaire responses. It can take up to 15 seconds to send a message

with a clear view of the sky under the best circumstances. It can take more than a minute if you're beneath trees with thin foliage.

The iPhone 14 series didn't come with satellite connectivity when it was launched. However, a software update made it available to customers in the United States and Canada in November. Germany, France, Ireland, and the United Kingdom did the same in December. Users in these nations are not required to pay a fee for the use of the technology for the first two years. As at the time of writing this guide, it is still being determined if it will be introduced to further markets or how much it will cost once the free trials have ended.

HOW TO SHARE YOUR LOCATION WITH FRIENDS

Find My is now compatible with the updated satellite connection. You can periodically check in and update your location to let your loved ones know where you are.

How to Use Find My to Share Your Location Via Satellite:

1. Access the Find My app.
2. Tap Start Sharing Location under the People tab.
3. Add the contacts you want to share with
4. If the iPhone lacks cellular service, you will receive instructions on connecting to a satellite.

The fact that your location has been provided by satellite will be indicated to receivers, letting them know that it won't update frequently.

How much will Emergency SOS via satellite cost?

The Emergency SOS via satellite feature of the iPhone 14 is free for two years beginning on the day the device is activated. The cost of the new emergency service is currently unknown once these two years have passed.

We hope you never need to utilize Apple's Emergency SOS via satellite capability, but it does sound like it could save your life in an emergency.

WHAT IS CRASH DETECTION?

I f it detects a serious traffic accident, your iPhone 14 can help alert emergency authorities and notify emergency contacts.

How does crash detection work in iPhone 14?

When your iPhone detects a serious auto accident, it will display an alarm if you don't cancel the automated emergency call within 20 seconds. If you are not responding, your iPhone will play an audio message to emergency services, letting them know that you have been in a serious accident and giving them your latitude and longitude along with a rough search radius.

When a crash is identified, the emergency calls placed via other ways won't be interrupted.

Suppose there is no Wi-Fi or cellular service nearby, and you are in a serious car accident. In that case, your iPhone will use Emergency SOS to try to contact emergency personnel via satellite if it is possible.

What kind of crashes can it detect?

Apple cautions that not all crashes may be caught by crash detection. You will naturally want to know when, how, and how well crash detection functions if you consider it important. But it's not like other phone features. Testing "find my phone" is not difficult when your phone is not missing. You can check to see if the feature is active and functional and how it functions. You cannot test how your iPhone or Apple Watch reacts by having a few vehicle accidents.

How to enable crash detection

By default, Crash Detection is turned on. In Settings > Emergency SOS, turn off Call After Severe Crash to stop receiving notifications and automatic emergency calls from Apple following a serious auto accident. Third-party apps that have been set up to monitor crashes on your device will still receive notifications.

HOW TO USE APPLE PAY

Use Apple Pay on your iPhone to make contactless payments.

You may use Apple Pay to make safe, contactless purchases in shops, restaurants, and other establishments by storing your credit, debit, and Apple Cash cards in the Wallet app on your iPhone.

Setting up Apple Pay

Add a debit, credit, or prepaid card to the Wallet app on your iPhone, Apple Watch, or other supported device to activate Apple Pay.

Apple Pay requires the following:

- a device1 that is compatible with the most recent iteration of iOS, iPad OS, watch OS, mac OS, etc.

- a supported card issued by a card issuer that participates.

- Your Apple device is logged in with an Apple ID.

    How to add a debit or credit card to Apple Watch

- Open the Apple Watch app on the attached iPhone.

- Choose Wallet & Apple Pay from the My Watch tab.

- Select Add Card.

- To add a new card, tap Debit or Credit Card.

- To add a previously used card, tap Previous Card.

- Select Continue.

- Verify your information with the bank or card issuer if necessary. Before allowing you to use your card with Apple Pay, they can require you to fill out a form or download an app.

Send money with messages

Use Messages to send or receive payments

- Tap the Apple Cash button in an i Message discussion then type the amount.

- If a dollar number is underlined in a message, tap it to establish the payment.

- Add a remark after tapping Pay (optional).

- By tapping the Send button and then entering your pass code, Face ID, or Touch ID, you may complete the transaction. You can use your debit card in Wallet to settle the bill if you don't have enough money in Apple Cash.

- A payment that hasn't been accepted can be canceled. Tap Cancel Payment after tapping the payment bubble.

- See Send or request payments with Apple Cash to send money using your wallet.

HOW TO MASTER IPHONE 14 PRO AND IPHONE 14 PRO MAX CAMERA

The brand-new iPhone 14 Pro and iPhone 14 Pro Max both include a Camera and a highly powerful app. Here's how to take the greatest pictures you can. The iPhone 14 Pro and iPhone 14 Pro Max have three rear cameras.

The gadget contains:

- A 48MP wide-angle primary camera.
- A 12MP ultra-wide lens.
- A 12MP telephoto lens.
- These other lenses.

Apple improved the camera this year with a revamped Photogenic Engine, Action Mode, and other features. Let's review the ever-expanding list of features available on Apple's most recent pro I Phones.

Volume buttons

Starting with the Camera app's controls, there are better options than tapping the screen. The phone can move with just one tap, accidentally distorting your picture.

Apple addresses this problem by enabling the volume buttons to serve as convenient shutter controls.

The volume up or down button will automatically take a picture when pressed. Holding either button initiates video capture. The video capture will end once the button has been released.

You can modify this behavior by going into Settings. You can enable burst capture by going to Settings > Camera.

When activated, holding the volume up button will take bursts of images instead. When you release the button, the images will no longer be taken.

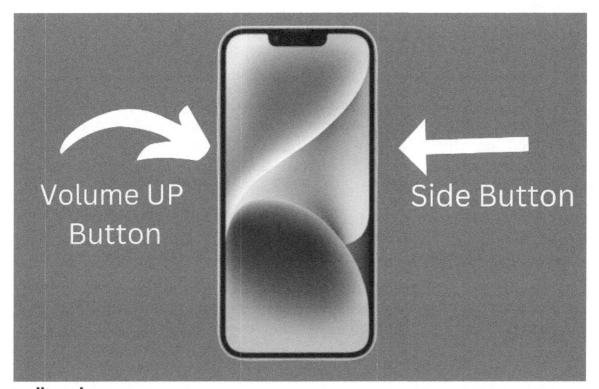

Controlling the zoom

Four optical-quality zoom levels are available on the iPhone 14 Pro and the iPhone 14 Pro Max. It exists five times using the ultra-wide lens, one time using the main wide lens, twice the main lens, and three times the telephoto lens.

The theory is that Apple can accomplish a 2X magnification without noticeably sacrificing quality by using the center 12MP from the larger 48MP sensor. If we were to go extremely scientific, there would be some quality loss because Apple cannot apply pixel binning when this occurs, which means the smaller pixels collect a little less light.

Apple Quick Take

Quick Take makes it simple to immediately capture images, movies, or burst shots. If you tap the white shutter button, it will take a picture for you; if you hold it, it will take a video, avoiding the need to switch to video mode.

As long as you keep your finger on the button, you can record a video. The recording stops when you let go. Don't hold down the shutter button if you want to record for a long time. To lock the camera in video mode, slide the shutter button to the right.

Bonus controls

Additional controls based on your shooting scenario are situated near the top of the phone, on either side of Dynamic Island. The flash icon, the icon for Night Mode, the RAW or Pro Res indicator, and the toggle for Live Photo may all be visible.

In addition, you may access more controls above the shutter by tapping the carrot arrow in the center of the screen. Among the icons available are those for the flash control (auto, off, on), Night Mode (appearing only in low light), Live Photo (on, off), Photographic Styles (choose one of five styles), aspect ratio (4:3, 1:1, 16:9), exposure compensation, timer, and filters.

Instantaneous image, video, or burst shot capturing is simple using Quick Take. The white shutter button will take a picture if you tap it; if you hold it down, it will record a video, avoiding the requirement to enter video mode.

As long as you keep your finger on the button, you can record a video. The recording stops when you let go. Don't hold down the shutter button if you want to record for a long time. To lock the camera in video mode, slide the shutter button to the right.

Video recording by iPhone 14

The iPhone 14 Pro comes with four video settings from Apple. There is ordinary video mode, time-lapse, slo-mo, and cinematic.

When you switch to video mode, the resolution and frame rate will be shown in the top-left corner. Tap one of the numbers to change. Depending on your video mode, several recording resolutions and frame rates will be offered based on what the phone can capture.

Apple improved Cinematic Mode this year to record video at up to 4K resolution on the iPhone 14 Pro. Before this, it could only be in HD 1080P. You can record at 24 to 30 frames per second when shooting in 4K.

iOS 16

How to customize/add widgets to your iOS 16 lock screen

You can add widgets to your iPhone lock screen and change any other part of the lock screen. It's a little different from what Apple has had in the past, but it shouldn't be too hard once you get used to it.

Step 1: Go to your iPhone's lock screen and press and hold on any space to open the lock screen menu. Here, tap Customize and then choose your lock screen as the thing you want to change.

Step 2: Look under the clock. You should see a message that says, "Add widgets." Tap on it to start the process of adding widgets.

Step 3: A widget picker that shows all of your iPhone's widgets will appear. This will include many of Apple's own apps and apps from other companies that have been updated to work with them.

Step 4: Choose the things you want. You can choose up to four, but how they are set up will depend on the type of widget you choose. You can take as many as four small widgets, two large widgets, or one large and two small ones.

Step 5: After you choose the widget you want, tap "Done" to finish. If you make a new lock screen, you will see Set as wallpaper pair. Tap and save.

There's no doubt that the widgets in iOS 16 could be better. Apple could let you customize them by putting more widget space on the lock screen than the current four, or it could make some of them more interactive, like live activities. Also, iOS 16 lock screen widgets make the new lock screen more customizable than ever since full widgets are just a swipe away.

How to Change Notification's Layout on Your Lock Screen

In iOS 16, Apple changed how the iPhone Lock Screen works. You can now add widgets and change the Lock Screen in more than one way. As part of its redesign, Apple changed how notifications look and gave users more ways to choose how they should be shown.

Now that the Lock Screen looks different, notifications come in from the bottom of the screen instead of the top. This makes the notification less obvious and keeps it from getting in the way of any widgets you have set up. You can see the notifications by swiping them up from the bottom of the screen, and you can hide them by swiping them down.

You can also change the layouts by pressing the list of notifications on the Lock Screen. Notifications can look like a list, a stack, or a number count that tells you how many are waiting for your attention. You can also use a new menu in Settings to make one of these layouts the default for your Lock Screen.

- ✪ Open the Settings app on your iPhone.
- ✪ Click "Notifications."
- ✪ Click Display As.
- ✪ Click one of the options (Count, Stack, or List) to choose it.

That's the end of the story. The change will take effect immediately, and all future notifications should appear on the Lock Screen based on the setting you chose.

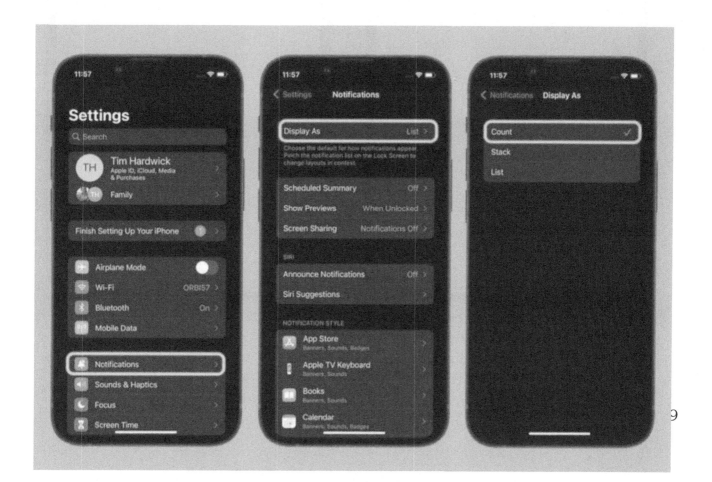

How to edit and unsend iPhone messages on iOS 16

To unsend the message—remember, you have two minutes—just long-press on the message and choose Undo Send. So, that's it, nothing too hard. But remember that the person you sent the message to will see that you unsent it, even though they won't be able to see it.

To edit a message, follow the same steps: long-press the message you want to change and then tap Edit. Remember, you can only do this five times in 15 minutes. Once your message is how you want it, tap the blue check, and the new version will be sent to your recipient.

And remember that the person you're sending it to can see the changes you've made. So, if you text your ex, "I still love you and I can't stop thinking about you," and then change it to "Hey, I have some of your mail," it could be an awkward conversation.

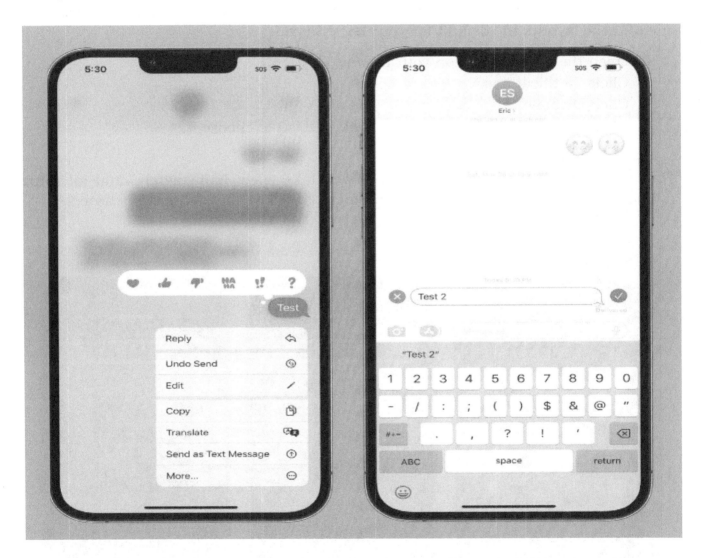

How to Recover Deleted Messages

Apple has made some wonderful changes to its stock Messages app in iOS 16. One of these is the ability to get back messages from conversations that you have deleted. Read on to find out how it works. In iOS 16, Apple's Messages app has a "Recently Deleted" section that shows all the text messages you've deleted in the last 30 days. So, if you accidentally delete a message and want to get it back, you can now do it right in the Messages app. Here's how to get back a message you accidentally deleted on an iPhone or iPad with iOS 16 or iPadOS 16.

- ✪ Go to the main Messages screen in the app to find filters like "Known Senders" and "Unknown Senders."
- ✪ Tap Deleted Recently.
- ✪ Tap the messages you want to get back until a blue check mark appears next to each one. (Note that each message shows how many days are left until it is deleted automatically.)
- ✪ Tap Recover in the bottom right corner of the screen. To confirm, tap Recover Message[s] in the pop-up. That's the end of the story. You can permanently delete messages from the same "Recently Deleted" screen by tapping them and then choosing "Delete" in the bottom-left corner. Keep in mind that it can take up to 40 days for Apple's servers to delete a message for good.

How to use iOS 16's Focus Filters

- ✪ Go to Settings > Focus > choose a Focus
- ✪ Swipe to the bottom and tap Add Filter
- ✪ Pick your filters and tap Add when you've customized them

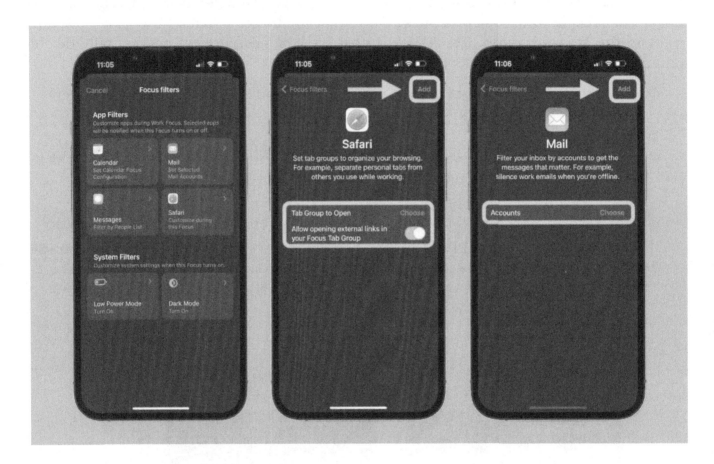

How to set up a schedule for iOS 16 Mail

You can now schedule emails to be sent later using iOS 16's Mail app. It's about time that iOS could let you schedule emails. Many of the best email apps for the iPhone let you choose when messages are sent. Apple's Mail app, on the other hand, has yet to be one of them. With iOS 16, you can send the email at a set time, so messages go out when you want them to.

With scheduled send, you can email your boss when you know they'll be checking their messages. Scheduled Send also lets you write emails whenever the mood strikes but send them at a better time.

The scheduled send feature in iOS 16 Mail is easy to use but hard to find. We will show you how to schedule an email in iOS 16 Mail so that every message you send gets there on time.

1. Write your message in iOS 16 Mail then tap and hold the Send button, which is the arrow next to the subject line in the upper right corner.

2. Then, a menu will pop up. You can send the message now, later tonight, or early in the next day's morning. There is also a Send Later option, which gives you more control over when to send the message. Choose one by tapping it. We'll use the Send Later option.

3. You'll see a calendar. Tap on the date you want to send the message or leave it on the current date if you want it to go out today.

4. Tap on the time to choose when you want your email to be sent. Use the rollers to set the hour and minute and tap Time Zone if you want to choose a different time zone than the one you're in by default.

5. Tap "Done" when the time is right.

Your message is now set to go out. You can see your email queue by going to the Mailboxes screen and tapping the Send Later button.

7. Tap the message and then tap edit to change the time you want to send an email. You can also delete the scheduled email by swiping left on the message and tapping the Delete button.

And that's all you have to do to send an email at a certain time in iOS 16 Mail. As you can see in our iOS 16 public beta hands-on, this is just one of the many changes Apple is making to the iPhone.

We have tips on using the new features in iOS 16, like changing a text message or taking it back in Messages. In iOS 16, you can also change how your iPhone's lock screen shows notifications or make it look different. You can also cut a person out of a photo, separate them from their background, and paste them into an email. You can choose to send that email later if you want to.

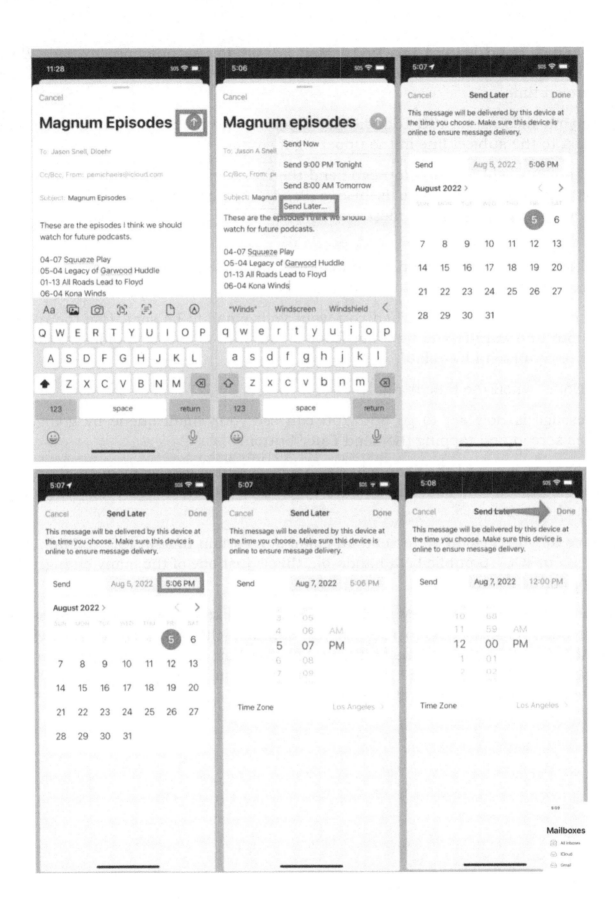

How to enable Lockdown Mode on an iPhone

1. Open the app for Settings.
2. Click on Privacy and Security.
3. Tap "Lockdown Mode" and "Turn on Lockdown Mode" under "Security."
4. Tap Lockdown Mode to turn it on.
5. Tap Turn On and Restart and then enter the passcode for your device.

When Lockdown Mode is on, you might get a message when an app or feature is limited, and Safari will show a banner that says Lockdown Mode is on.

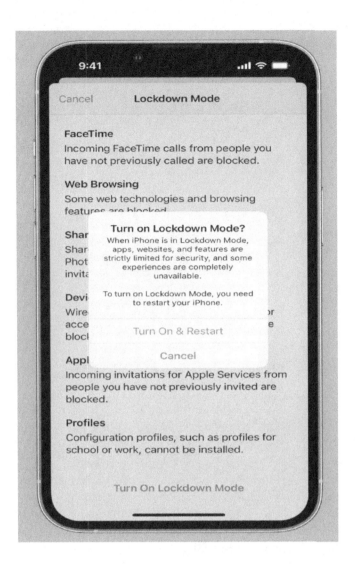

How to share Safari Tab Groups

To share tabbed groups on an iPhone, open Safari, tap the square icon in the bottom right corner to access your open tabs, choose the Tab Group you want to share from the drop-down menu, and then look for and hit the newly arriving Share icon in the top right corner of the page. You'll then be asked whom you want to share the collection with, and you'll be prompted to send them a message inviting them to your group.

You can share it with numerous individuals, and once you've shared a collection, you can check who shared it by tapping the new user profile icon at the top-right of the tab browser interface.

How to Display the iPhone's Battery Percentage in iOS 16

The battery percentage option is off by default; therefore, you must enable it manually.

1. Open the iPhone Settings app.
2. Scroll to the bottom and select Battery.
3. Switch the battery percentage option to On. Toggle the setting to Off to disable the feature.

You'll now see your iPhone's battery percentage inside the battery symbol, as opposed to the previous method where the battery percentage was to the left of the symbol.

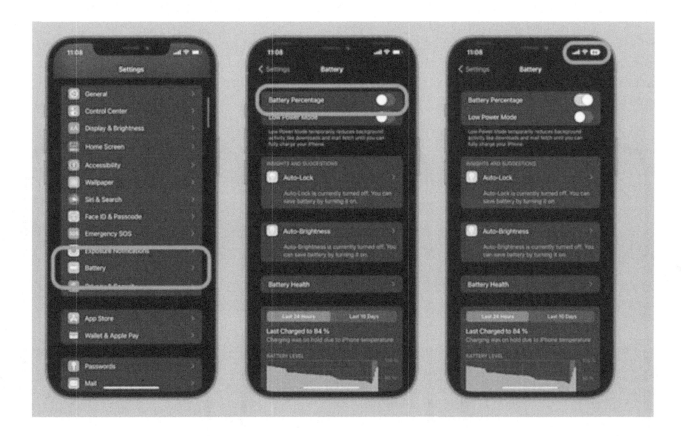

How to set up a route with multiple stops in Apple Maps on iOS 16

1. Open the Maps app.
2. Tap or click the Search Maps field then enter an address or a search term to look for a place.
3. Choose your place to go from the list of search results.
4. Tap the blue "Directions" button. The button will have an icon that matches the mode of transportation you chose: driving, walking, public transportation, cycling, or, if available, a ride-sharing app. In the next step, you can change this setting for this route if you want to.
5. You will be given a list of different paths you can take. Here is where the new "Add Stop" button can be found.
6. When you tap or click "Add Stop," you'll be asked to type in or look for another stop.
7. Repeat this process until your route is completely calculated then tap the green "Go" button to start navigating.

The stop is added to the list of places to go so that it will go at the end of the route. From a practical point of view, this might take some time to get used to. Before you start driving, you can easily change the order of your stops by dragging the bars on the right.

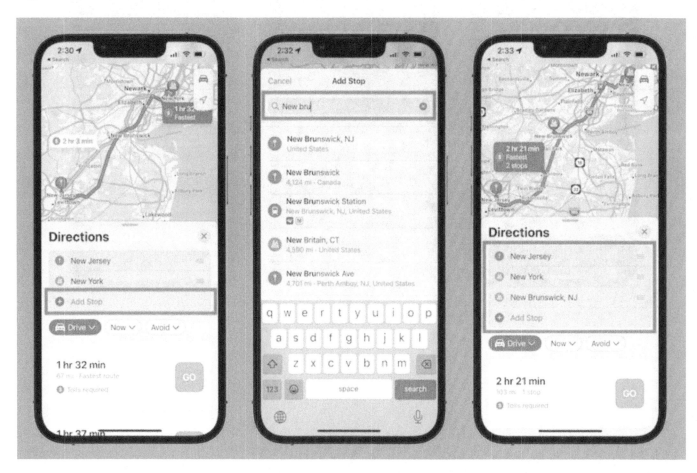

Track your fitness app without a stopwatch

1. Open the app for Fitness
2. Click on the picture of you.
3. Press Details About Health
4. Enter your details
5. Return to the main screen and click on the circles.
6. Check your numbers.

How to use Safety Check on an iPhone in iOS 16

1. Open the Settings app on iOS 16
2. Drag down the screen and tap Privacy & Security.
3. Slide down and click on Safety Check.
4. Now you can use Emergency Reset or Manage Sharing & Access. Face ID/Touch ID or passcode is needed.
5. Emergency Reset will immediately reset access for all people and apps and help you check your account security.
6. Manage Sharing & Access will let you choose which people and apps can access your information and review your account security.
7. Here's how to use it: iPhone Safety Check in iOS 16:

8. Apple says to use the Emergency Reset Safety Check for iPhone option if you think "your personal safety is at risk."

Apple won't tell anyone you were sharing with that you've stopped, but they may notice that sharing has stopped. If you don't want to use the feature, tap Cancel or Quick Exit at the top of your screen.

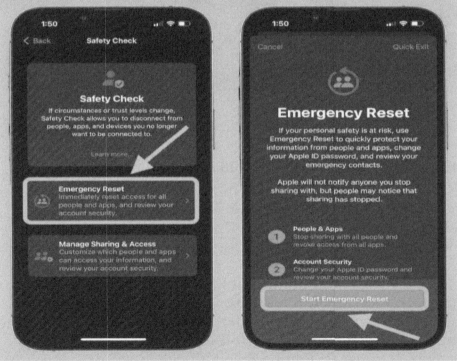

How to create image cutouts in iOS 16

One of the best things about iOS 16 is the lock screen, but you can also easily cut out people from photos, which is a fun but less obvious addition. It's easy to separate a subject from its background, share the results quickly, or copy and paste them into other programs. What usually takes a long time and a lot of work to do with photo editing software can be done quickly and easily with iOS 16. It's surprisingly good, and you can share the results easily with your group chat to make it even more fun. How does it work?

I did these steps on an iPhone 11 with iOS 16.0.2 and an iPhone 14 Pro with iOS 16.1 developer beta 3.

1. Open the Photos app from Apple.
2. Find a picture in your camera roll with a subject you want to cut out, such as a person, pet, or object that stands out.
3. Press down on the subject for a long time until you feel a buzz and see a white border around the subject.

4. If you copy the image, you can paste it anywhere that lets you paste a PNG image file, like Google search, Messages, or even a note. If you tap Share, you'll see options like "AirDrop," "Save to Camera Roll," "Assign to Contact," and "Add to New Quick Note."

That's it! Here are a few tips to help you get the best results possible:

- ✪ You can use any photo in your camera roll; it doesn't have to be one you took with your iPhone.
- ✪ It works best with people, pets, or things that are easy to point out.
- ✪ If you press an image for a long time and nothing happens, try again by pressing a different part of the image. If more than one try fails, there are better choices for image cutout.
- ✪ Tap either Copy or Share from here.

How to keep track of medications on an iPhone using iOS 16

✪ Open the Health app on an iPhone running iOS 16 and
✪ Select the Browse tab at the bottom right of the screen.
✪ Select Medications then Add a Medication.
✪ You can scan your medicine with your camera or type it in by hand (Apple says scanning will be limited to US users for now)
✪ Follow the directions to set reminders and do other things.
✪ Go to the Health app again > Browse > Medications to record what you've taken and more at any time.

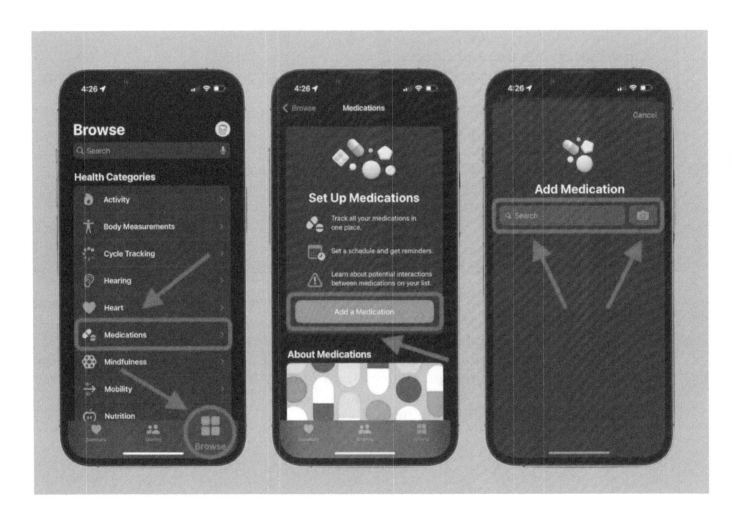

Here's how it looks on an iPhone to keep track of medicines:

The Health app will show a list of possible matches if there isn't an exact match.

Once you've added medications, vitamins, or supplements to your iPhone, you can set a reminder for when to take them and how often. Even the shape and color can be changed.

Adding a nickname/notes and deciding if you want to see possible interactions with alcohol, tobacco, and marijuana are the last steps to tracking medications on an iPhone. Interactions between medications will be shown automatically.

Here's what it looks like when you start tracking medications on your iPhone. You can tap on it to record whether you skipped or took it, among other things.

As shown above, from this main screen (Health app > Browse > Medications), you can add more medications, export your list as a PDF, and more.

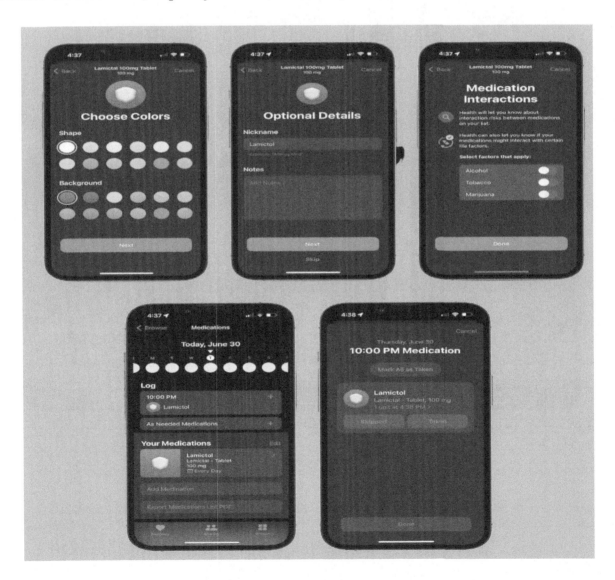

How to make iOS 16 use the haptic keyboard

Step 1: Open Settings.

Step 2: Click on Sounds and Touch.

Step 3: Go to the bottom of the page and tap on Keyboard feedback.

Step 4: To turn on Haptic, tap on it.

Step 5: You can also turn on the Sound option if you want each keypress to make a sound.

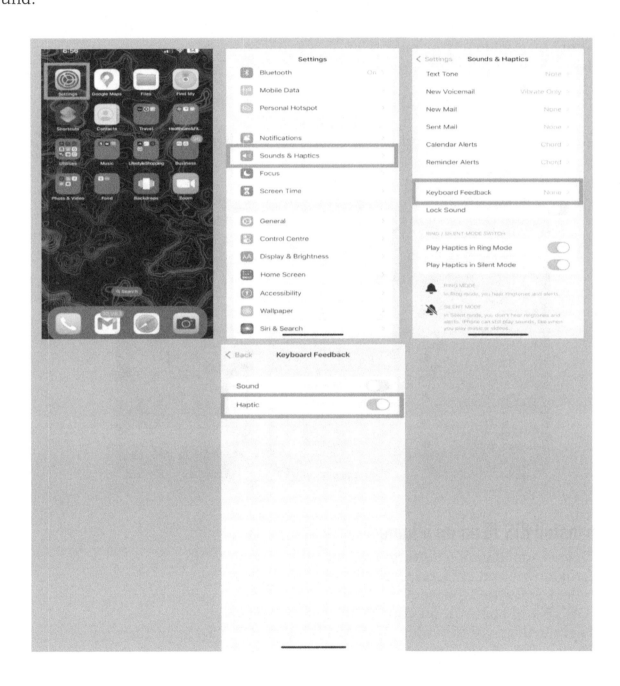

iCloud Shared Photo Library

Turn on the iCloud Shared Photo Library

Open the Settings app on your iPhone (or iPad on iPadOS 16.1)

1. Drag down and select Photos.
2. Tap Set Up Now under Shared Library then tap Get Started
3. Pick whether you want to invite people now or later.
4. Pick what you want to be part of the Shared Library.
5. If you want to, you can look at the Shared Library.
6. Send invitations for people to join. Choose whether you want the Camera to share automatically or manually.
7. Tap "Done," and you're done!

Note: It may take a while to finish and work perfectly.

You can also change your settings anytime by going to Settings > Photos > Shared Library. On an iPhone, this looks like this:

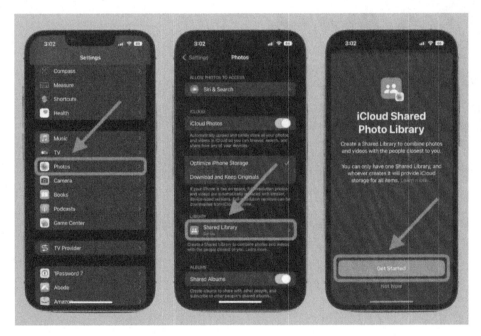

How to install iOS 16 on an iPhone

1. Open your iPhone's Settings app.
2. Drag down and click on General.
3. Select Software Update.
4. iOS 16 will show up on your device when it's ready.
5. Look at the bottom under "Also Available" if iOS 15.7 is at the top.
6. Under iOS 16, tap Download and Install.
7. Follow the on-screen instructions to finish installing.

Here's what it looks like to install iOS 16 on an iPhone: If your iPhone or iPad doesn't have enough space to install iOS 16, it will ask you if you want to automatically and temporarily delete some content.

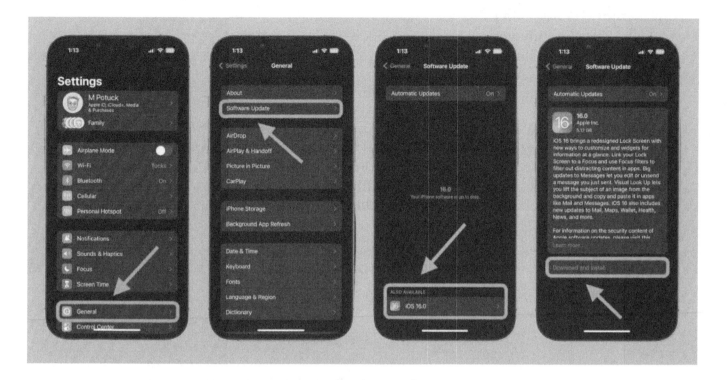

How to customize the iPhone lock screen, wallpaper font, color, and widget

How to create a custom lock screen for your iPhone

- ✪ When the Customize button displays at the bottom of the screen, touch and hold the Lock Screen.
- ✪ Touch and hold the Lock Screen once again then input your passcode if the Customize button isn't visible.
- ✪ Tap the "Add New" button at the bottom of the screen.
- ✪ The gallery of lock screen wallpapers emerges.
- ✪ To set one of the available wallpapers as your lock screen, tap it.
- ✪ You can swipe left or right on several wallpaper options to experiment with various color filters, patterns, and typefaces that go well together.
- ✪ Do one of the following after tapping Add:
- ✪ Select whether to display the wallpaper separately on the Home Screen and Lock Screen: Then select Set as Wallpaper Pair.
- ✪ Make additional adjustments to the Home Screen: Click on Customize Home Screen. To alter the wallpaper's color, choose a color, choose Photo on Rectangle to utilize a specific photo, or choose Blur to make the background less noticeable so that the apps stand out.

How to edit the lock screen

- ✪ You can modify your lock screen once you've created it.
- ✪ When the Customize button displays at the bottom of the screen, touch and hold the Lock Screen.
- ✪ Then touch the Add New button after swiping to the Lock Screen you wish to alter.
- ✪ Attempt one of the following:
- ✪ Pick a wall covering: Tap a button at the top of the screen, or select an option from one of the categories (such as Featured, Suggested Photos, or Photo Shuffle) (Photos, People, Photo Shuffle, Emoji, or Weather). See Customize your Lock Screen photo if you want to add a picture to your lock screen.

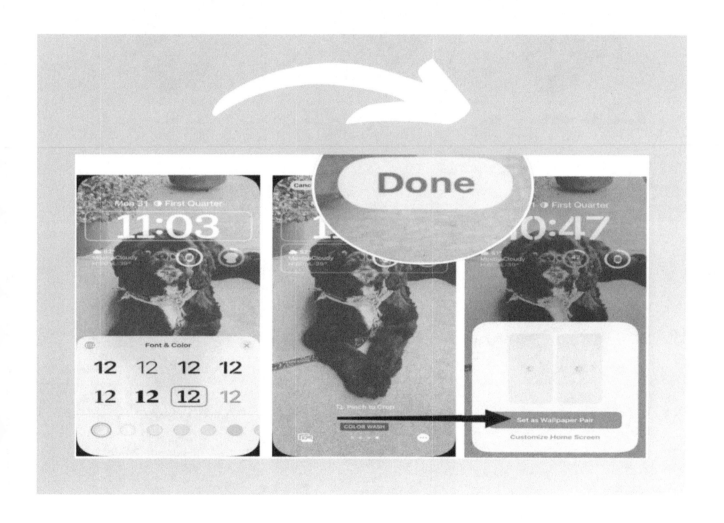

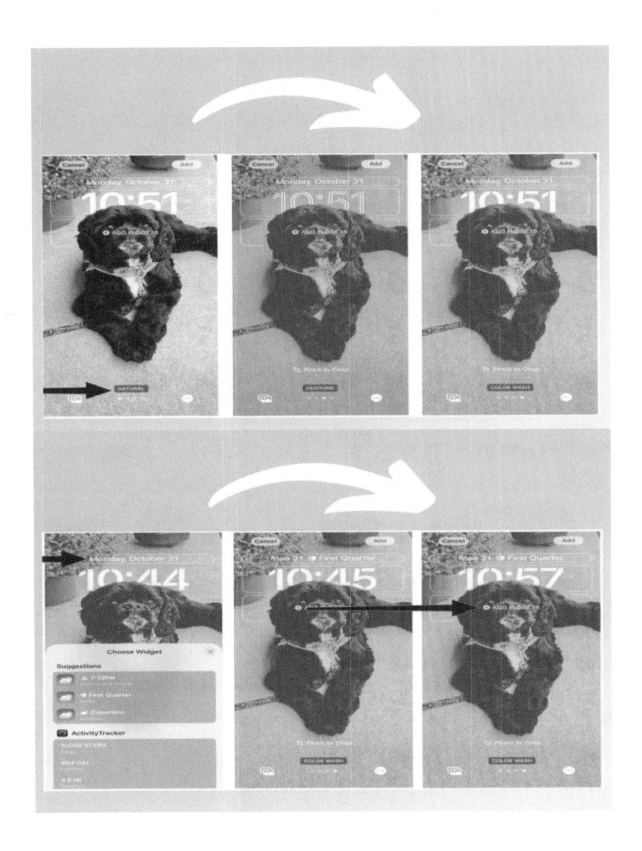

How to delete a lock screen

✪ Lock Screens that are no longer required can be deleted.

✪ When the Customize button displays at the bottom of the screen, touch and hold the Lock Screen.

✪ To erase a Lock Screen, swipe to reveal it then swipe up to see the Trash icon.

Swipe out the lock screen on the doors

✪ You can make several personalized Lock Screens and change them during the day. Switching from one Lock Screen to another also changes your Focus if a Lock Screen is linked to a particular Focus.

✪ When the Customize button displays at the bottom of the screen, touch and hold the Lock Screen.

✪ Tap the Lock Screen you want to use after swiping on it.

How to add a photo to the iPhone lock screen

✪ You can move the photo, alter the photo's appearance, and carry out other options if you select it for your Lock Screen.

✪ Reposition your image by pinching open to zoom in, dragging it with two fingers to move it, and then pinching closed to zoom out.

✪ Modifying the image's style: To experiment with several photo styles with complimentary typefaces and color effects, swipe left or right.

✪ Create a multilayered effect by tapping the "More" button at the bottom right and selecting Depth Effect if you have photo-supporting layerings, such as one with people, animals, or the sky.

✪ **Set the shuffle frequency:** If you select Photo Shuffle, you may preview the photos by hitting the Browse button and changing the shuffle frequency by tapping the "More" button and making a choice underneath Shuffle Frequency.

How to add widgets

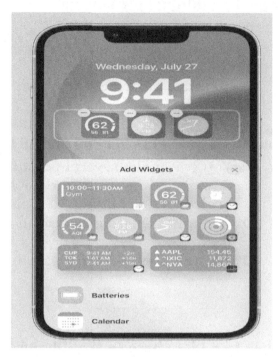

- ✪ You can add the widgets to your lock screen to quickly see things like the current temperature, how much battery life you have left, or upcoming events on your calendar.
- ✪ After touching and holding the Lock Screen until Customize button appears at the bottom of the screen, tap Customize.
- ✪ Tap the box beneath the time to add widgets to your Lock Screen.
- ✪ You can add widgets by tapping or dragging them.
- ✪ If there isn't enough space for a new widget, you can remove an existing widget to make way for a new one by tapping the Remove Widget button.

How to add Focus

- ✪ Focus reduces distractions so you can focus on a task. You can configure a Focus to allow only certain alerts or to temporarily quiet all notifications (ones that apply to your task, for example). Your Focus settings are applied when you utilize the Lock Screen you've linked to it.
- ✪ When the Customize button displays at the bottom of the screen, touch and hold the Lock Screen.
- ✪ If you want to see the Focus options, such as Do Not Disturb, Personal, Sleep, and Work, tap Focus near the bottom of the Wallpaper.
- ✪ After choosing a Focus, press the Close button.

How to customize iPhone 14 home screen

Change wallpaper

- ✪ Click on the Settings app from the home screen.
- ✪ After scrolling to and choosing Wallpaper, choose Add New Wallpaper.
- ✪ Locate the preferred Wallpaper and click it.
- ✪ After making any necessary wallpaper settings, choose Add. To set a wallpaper pair, select Set.

<u>Add widgets to the home screen.</u>

- ✪ On the home screen, click and hold a free area until the apps tremble then choose the Add icon. Pick the widget of your choice.
- ✪ Swipe to the left or right to find the layout and size of the widget you want then click Add Widget.
- ✪ Choose the required widget and drag it to the desired spot. When done, click Done.

<u>Use Smart Stack</u>

- ✪ Based on the apps you use the most frequently, a Smart Stack of widgets is intelligently curated. With Smart Stack, the appropriate widget will automatically appear at the appropriate time of day based on your app activity. Select and hold an empty area on your home screen until the apps begin to shake then select the Add icon, scroll to, and choose Smart Stack.
- ✪ Swipe left or right until the layout and size are as desired then choose Add Widget.
- ✪ To scroll across widgets on the Smart Stack widget, swipe up or down.

<u>Access App Library</u>

Your apps are all automatically categorized by the App Library into a single, intuitive display. Thanks to category-based app sorting, your most frequently used apps are always a swipe away. Swipe left from the home screen on the right to open the App Library then choose the required app.

Other ios16 tricks

<u>The Best iOS 16 Tips & Tricks for 2022 are listed below:</u>

Retrieve just-deleted chats from the message
With iOS 16, you can retrieve deleted messages if you have a habit of deleting them in the Message app and need them later. The Recently Removed section in the Message app itself receives the messages that have been deleted. Here's how to open Message's Recently Deleted folder:

- ✪ Launch the messaging app.
- ✪ In the top left corner, tap the Edit button.
- ✪ A recent deletion folder should be chosen.
- ✪ Choose the chat from which you wish to retrieve the texts.

Delete contacts quickly
A contact on an iPhone was difficult to remove with earlier iOS versions. You needed to access the contact's specific information, choose Edit, and then scroll down to Delete Contact. With iOS 16, Apple has finally listened to customer requests and made it

simpler to delete contacts. You must tap and hold on to a certain contact to remove it from your list.

Turn off the Lock to End the Call.
Pressing the lock button ends the connection, one of the quirks that iPhone users (especially new users who migrate from Android) lament. With this iOS 16 hack, you can finally turn off the "Lock to End Call" feature.

- ✪ Launch the Settings app.
- ✪ Click on Accessibility.
- ✪ Choose Touch.
- ✪ Switch the Lock to End Call off.

Use Siri to end the call.
On iOS 16, you can tell your virtual assistant, Siri, to hang up a call for you instead of tapping the screen's end call button or the device's side button. Say, "Hey Siri, hang up" while you are on a call, and the virtual assistant will do it for you. Although it is a practical and hands-free method of ending a call, it is advised to use it with caution because the person on the other end can hear your command to Siri.

Use Siri to restart your iPhone.
Restarting your iPhone using Siri is another iOS 16 technique. Say "reboot" or "restart" to your assistant when you hear the "Hey Siri" response. Respond "Yes" or "No" when Siri requests confirmation, and that's it! It can also be done remotely using Siri since you don't even need to unlock your device to finish the task.

Use Siri to dictate emojis.
Previously, you could use Siri to dictate texts, but if you're the kind who believes that an emoji is worth a thousand words, Siri was only a little use to you. Siri will type "heart emoji" in the text if you dictate a heart-shaped message to your sweetheart. It severely deflates the mood. Now that Siri is more knowledgeable about spoken emojis thanks to iOS 16, texting the assistant "heart emoji" will result in a lovely red heart appearing immediately in your message.

Do not ask Siri to confirm
Previously, Siri would request your approval before sending a message to a contact you had dictated. The advantage is that you can verify whether Siri successfully transcribed your message. However, it involves an extra step and requires more time. You may disable it and use Siri to send messages automatically with this iOS 16 hack. This is how:

Launch the Settings app.
- ✪ Activate Siri & Search.
- ✪ Select Send Messages Automatically.
- ✪ Toggle Messages Will Be Sent Automatically on.

Extend Siri's listening time

Let's talk about an accessibility feature Apple unveiled in May, a month before its iOS 16 preview while talking about Siri updates. When Apple announced that a variety of accessibility improvements would be added to the iPhone, one of the features that attracted my attention was the option to modify Siri's pause time.

If you change the pause time, the digital assistant will wait longer before responding to your request. This iOS 16 feature is great for people who have speech impairments, but you'll also like it if you frequently rush to finish a Siri demand before the assistant responds.

- ✪ Access Settings.
- ✪ Click on Accessibility.
- ✪ Touch Siri.

Pinned Tabs in Safari

In iOS 16 and iPad OS 16, pinned tabs are now available and can be used in tab groups. Pinning is a choice if you have tabs in your Tab Group that you wish to keep open all the time. Open the Tab Group, long-press the URL bar in Safari, and select "Pin Tab" to pin a tab to the Tab Group in iOS 16.

Combine Multiple Contacts

This iOS 16 tip is for you if your Contacts app is clogged with duplicates that you still need to remove. The update comes with a function specifically designed to merge duplicate contacts, ending issues with multiple contacts.

If you have multiple contact cards for the same person in the iOS 16 Contacts app, the app will automatically identify the duplicate and notify you that duplicates have been identified. The Contacts app can combine all duplicate contact cards into one if you tap the "Duplicates Found" interface, located at the top of the app.

HEALTH APP WITH IOS16

The health app was created to organize your important health information and make it available in one secure area. Improvements to Health Sharing, a new way to manage, understand and track your medicines, and significant changes to Sleep are all included in iOS 16.

Monitoring Medication

✪ Open the Health app on an iPhone that is running iOS 16
✪ In the lower right corner, select Browse.
✪ Tap Medications then select Add a Medication.
✪ Scan your medication with your phone's camera or manually type it in (Apple says scanning will be limited to US users for now)
✪ Observe the instructions to add reminders and more.
✪ Re-open the Health app > To record what you've taken and more, go to the Browse tab > Medications at any time.

Medication logging

✪ You can log a medication from the Health app on your iPhone or the Medications app on your Apple Watch.
✪ Activate the Health app.
✪ Following Browse, select Medications.
✪ Select the drug then tap the plus sign (+). You can tap the medication then tap Log for As-Needed drugs.
✪ Before tapping Done, tap Skipped or Taken.

Apple Watch medication app

- ✪ Activate the Medication app., the Apple Watch app's medication management icon.
- ✪ Then select a drug.
- ✪ To log the medication, select Log as Taken. If you didn't take the medication, you could tap Skipped, then Done.
- ✪ If you only take the drug when needed, you can scroll to it in the Your Medications section, press it, and tap Log.

Medication interactions

- ✪ If you live in the US, you can check to see if any drugs you take have a Moderate, Serious, or Critical drug interaction. Open the Health app on your iPhone.
- ✪ Following Browse, select Medications.
- ✪ Tap Drug Interactions after scrolling to Your Medications.
- ✪ To get a summary of the interaction information, choose an interaction.
- ✪ To modify the drug factors the health app looks for, tap Edit next to Interaction Factors.

Medication side effects

If there are any available adverse effects for any medication you select from your Health app list, you can view them by selecting the "Side Effects" button. Elsevier, a Dutch corporation that specializes in medical content and publishes the Annual ScienceDirect book series on drug side effects, provides the data, albeit not all pharmaceuticals have recorded side effects.

Export Medications

A basic PDF of your medications can be created by selecting the "Export Medications List PDF" option in the health app's Medications section. You can give a doctor this list.

Sleep Features

Watch OS 9 and iOS 16 add additional features for people who wear the Apple Watch to bed to track their Sleep. You may determine how well you slept by comparing the categories of Awake, REM, Core (light), and Deep Sleep in the sleep data collected by the Apple Watch.

When you go to bed, how long it takes you to fall asleep, how frequently you wake up, and how much time you spend in REM, Core, and Deep Sleep are all monitored by the Apple Watch.

Awake - You might have periods of partial waking during a sleep session. It's common for people to wake up occasionally. You may go back to sleep after waking up and forget about it.

REM: This sleep state may aid in memory and learning. Your muscles are most relaxed, and your eyes move quickly from side to side. Additionally, most of your dreams happen at this time.

Core: This stage, sometimes called light Sleep, is equally as significant as the others. This phase frequently corresponds to the majority of your nightly Sleep. During this phase, brain waves that could be crucial for cognition occur.

Health Sharing Invitations

In iOS 16, you can invite family members to share their data with you, making it simple to monitor the health data of children or senior family members. One can choose which information to share with you after accepting an invitation.

SETTING UP FITNESS APP WITH IOS16

Issues and Troubleshooting

The Fitness app on iPhone may not function for a variety of reasons. Let's look at the most typical ones.

- ✪ You haven't activated fitness tracking.
- ✪ Your Apple Watch and iPhone were not properly connected.
- ✪ You might be using an out-of-date watch OS version.
- ✪ Your iPhone might have a system-level problem.

There are several reasons why the iPhone's Fitness app might not work. Let's examine the most prevalent ones.

- ✪ You haven't turned on the fitness monitoring.
- ✪ Your Apple Watch and iPhone may not have been properly connected.
- ✪ You may need an updated watch OS version.
- ✪ Your iPhone may have a system-level issue.
- ✪ Open the Watch app and navigate to Privacy to be sure. If the toggles for Heart Rate and Fitness Tracking are off, turn them on.

Update software
- ✪ Your smartphone can be backed up via iCloud or your computer.
- ✪ Connect your device to a Wi-Fi network and a power supply to access the internet.
- ✪ Once you've navigated to Settings > General, choose Software Update.
- ✪ If you have a choice among several software update options, select it.

○ Choosing "Install Now." As an alternative, if you see Download and Install, click it to download the update, then, after entering your passcode, press Install Now. Learn what steps to take if you need to remember your passcode.

Set automatic updates to on
○ Access Software Update by going to Settings > General.
○ Turn on Download iOS Updates by tapping Automatic Updates.
○ Install iOS Updates must be enabled. The most recent iOS or iPad OS version will be installed on your device automatically.

Obtain and install Rapid Security Responses
○ Before becoming included in other enhancements in a subsequent software release, Rapid Security Responses deliver significant security improvements more quickly.
○ To automatically receive Rapid Security Responses:
○ Access Software Update by going to Settings > General.
○ Select Automatic Updates.
○ Check to see if "Security Responses & System Files" is enabled.

If a Rapid Security Response needs to be removed:
○ Activate Settings > General > About.
○ Select iOS Version.
○ Remove Security Response by tapping.
○ Rapid Security Response can be reinstalled later, or a regular software update can be installed permanently.

Check motion calibration and distance setting.
○ Check these settings on your iPhone to ensure that your Apple Watch can get the data it requires:
○ Open the Settings app on your iPhone.
○ Select Location Services under Privacy & Security.
○ Make sure Location Services are activated.
○ After descending, tap System Services.
○ Check to see if Motion Calibration & Distance is activated.
○ Follow the next few steps.

Set your Apple Watch's calibration
○ Wear your Apple Watch and head to a level, large outdoor area with good GPS reception and clear skies.

- If you own a Series 2 or later model, all you need is your Apple Watch. If you have an Apple Watch Series 1 or earlier for GPS, bring your iPhone with you. You can carry your iPhone in your hand, on your arm, or around your waist.
- Once the Workout app has been launched, select Outdoor Walk or Outdoor Run. To set a goal before starting, tap.
- Pace yourself for around 20 minutes while you run or stroll.

You can spread out the 20 minutes over numerous outdoor training sessions if you are pressed for time. If you work out at different intensities, calibrate for 20 minutes at each walking or running pace.

When you exercise outside using the abovementioned methods, your Apple Watch calibrates the accelerometer by calculating your stride length at various speeds. Calibration can help with calorie estimations in the Workout app and calorie, distance, Move, and Exercise estimates in the Activity app.

Boost the accuracy of your workouts and activities

Your Apple Watch uses personal data, including your height, weight, gender, and age, to determine how many calories you burn daily. Find out how to update your data.

Calibration data reset

The steps below to reset your calibration data:

- Get your iPhone's Watch app open.
- Then, choose Privacy > Reset Fitness Calibration Data from the My Watch page.

TRICKS TO BOOST BATTERY LIFE

An unexpected power depletion issue is one of the worst problems that might happen to your iPhone after installing iOS 16. Even after the most recent software update, this problem may still exist. Even though iOS 16 has only been officially available for a short while, there have already been numerous claims of battery drain from the general public.

So, are you too using iOS 16? Do you frequently find that your iPhone battery is dead before the day is over? This in-depth article offers more than 12 tricks, adjustments, and ways to easily extend the battery life of an iPhone running iOS 16 or any other version. Let's look at it.

- ✪ Deactivate Location Services
- ✪ Take unneeded widgets off your home screen.
- ✪ Employ dark mode.
- ✪ Turn off the background app, refreshing
- ✪ enable automatic brightness
- ✪ Stop using unused apps
- ✪ Regularly restart your iPhone.
- ✪ Lay your iPhone on its side.
- ✪ Turn off the automatic app and download updates.
- ✪ Activate optimum battery charging
- ✪ and activate Low Power Mode.

✪ The iOS version on iPhone updated

How to save battery

✪ When not in use, turn off Bluetooth and Wi-Fi. Use the Settings app rather than the Control Center whenever possible.

✪ 30-second auto-lock interval setting: Select 30 Seconds under Auto-Lock in the Display & Brightness section of the Settings app.

✪ Reduce Motion is activated: Open the Settings app, select Accessibility, Motion, and activate the Reduce Motion toggle.

Deactivate dynamic wallpaper

✪ Use static or live wallpaper at your peril. Use a still photo instead.

✪ In places with weak signals, such as on trains, use Airplane mode.

✪ Switch off iPhone's "Raise to Wake" feature: Turn off Raise to Wake in the Display & Brightness section of the Settings app.

WAYS TO SELL OR TRADE OLD IPHONE

There are several ways to sell or trade-in an old iPhone:

Online marketplaces:

You can sell your old iPhone on online marketplaces such as eBay or Amazon.

Trade-in programs:

Many retailers, such as Apple, Best Buy, and Walmart, have trade-in programs that allow you to trade in your old iPhone for a gift card or credit toward a new device.

Recycling programs:

Some companies, such as Gazelle, will pay you for your old iPhone and then recycle it.

Social media:

You can also use the social media platforms, such as Facebook Marketplace, to sell your old iPhone to people in your local area.

Specialized websites:

Websites such as SellYourMac.com, uSell.com, and NextWorth.com allow you to sell your iPhone directly to them.

Selling to a friend or family member:

If you know someone who needs an iPhone, you can sell it to them at a discounted price.

Dear reader, if you're reading this sentence, you probably haven't carefully read the description of this book on the Amazon page, where the link to buy the COLOR version is clearly indicated at the end.

But don't worry, I have a surprise for you:

SCAN THE QR CODE TO DOWNLOAD AND ENJOY THE COLOR VERSION!

IPHONE 14 PROTECTION

The best covers for Apple's most recent models are listed here, and they are all Mag-Safe-compatible.

Speck cases

Many of Speck's iPhone 13 case designs, like the Presidio Perfect-Clear and Perfect-Clear Grips and the Presidio2 Pro, have been moved over to the iPhone 14. The Candyshell Pro case from Speck is still the least expensive at about $25, but it doesn't have MagSafe. Cases with MagSafe built-in cost about $50. Also, all Speck cases are 40% off right now, so you can get one for a lot less.

Speck's Microban antibacterial protection and best drop protection (13 to 16 feet, depending on the model) are features shared by all the new models. It's No matter which model you choose. Speck often gives discounts to new customers.

Spigen Ultra Hybrid MagFit

MagSafe cases can be rather expensive, but Spigen's Ultra Hybrid MagFit is a nice deal at roughly $25 if you're searching for a cheap transparent MagSafe case for your new iPhone 14 series phone. There are various additional iPhone 14 case alternatives from Spigen and a non-Mag version of this case that costs around $8 cheaper.

OtterBox Symmetry Series Plus

OtterBox gained popularity for their incredibly durable Defender series cases, but nowadays, most users choose something more lightweight. I enjoy the Symmetry Series Plus cases since they have MagSafe and come in various colors, including the new

Euphoria color. It's crucial that the Symmetry Plus Pop case supports wireless charging and has a PopSockets PopGrip. On its website, OtterBox provides a discount for first-time customers.

Case-Mate Blox

Case-Mate offers several striking cases for the many iPhone 14 variants, but my favorite is the squarish Blox case. It provides a wide range of color options, supports MagSafe, has decent corner protection, and is competitively priced. It's worth looking at the other colors, such as rainbow frosting and the clear variant, which I'm showcasing on a deep purple iPhone 14 Pro.

Caseology Capella Mag Clear

The NanoPop, Parallax, Skyfall, and Vault are just a few respectable low-budget cases that Caseology has produced in the past. One of the more affordable transparent MagSafe cases we've seen is its new Capella Mag Clear.

Lupa Legacy Wallet Case

A few different wallet cases are available from Lupa Legacy for the iPhone 14 versions. The folio version, which comes in various color options and has a magnetic clasp, is what I prefer. It offers reasonable corner protection against drops and includes three slots on the interior of the cover where credit cards or cash can be stored. The synthetic leather seems respectable (the case looks more expensive than it is, which is a good thing).

The case's lack of a MagSafe and inability to function as a kickstand are its sole downsides. However, a wireless charging pad works perfectly for charging your phone.

Element Case Special Ops X5 MagSafe case

The $250 Back Ops X5 case from Element Case is made for the iPhone 14, but there are more cheap options as well. X5 Special Ops To help keep your screen and cameras from cracking, the MagSafe case has thicker corner protection and raised edges on the back and front. It has good side grips, too. The only issue I have with it is that I wish there were more color choices. It's a strong argument.

Cyrill cases

Although they cost a little more, Cyrill's cases are undoubtedly more fashionable than Spigen's. Due to its faux "vegan" leather back, the Kajuk Mag ($27) is also rather good, and I prefer the UltraColor Mag ($25). If you prefer a little flash in your iPhone case, the clear Shine Mag ($30) cases are nicely designed with "a touch of shine."

INDEX

Made in the USA
Las Vegas, NV
20 November 2023